AF449203

Effective Environmental Management
in Developing Countries

Other books by IDE-JETRO also published by Palgrave Macmillan

Masahisa Fujita (*editor*)
REGIONAL INTEGRATION IN EAST ASIA

Daisuke Hiratsuka (*editor*)
EAST ASIA'S DE FACTO ECONOMIC INTEGRATION

Akifumi Kuchiki and Masatsugu Tsuji (*editors*)
INDUSTRIAL CULTURES IN ASIA

Mayumi Murayama (*editor*)
GENDER AND DEVELOPMENT

Nobuhiro Okamoto and Takeo Ihara (*editors*)
SPATIAL STRUCTURE AND REGIONAL DEVELOPMENT IN CHINA

Tadayoshi Terao and Kenji Otsuka (*editors*)
DEVELOPMENT OF ENVIRONMENTAL POLICY IN JAPAN
AND ASIAN COUNTRIES

Mariko Watanabe
RECOVERING FINANCIAL SYSTEMS

Effective Environmental Management in Developing Countries

Assessing Social Capacity Development

Edited by

Shunji Matsuoka

First published 2007 by
PALGRAVE MACMILLAN
Houndmills, Basingstoke, Hampshire RG21 6XS and
175 Fifth Avenue, New York, N.Y. 10010
Companies and representatives throughout the world.

PALGRAVE MACMILLAN is the global academic imprint of the Palgrave Macmillan division of St. Martin's Press, LLC and of Palgrave Macmillan Ltd. Macmillan® is a registered trademark in the United States, United Kingdom and other countries. Palgrave is a registered trademark in the European Union and other countries.

ISBN-13: 978–0–230–54276–1
ISBN-10: 0–230–54276–X

This book is printed on paper suitable for recycling and made from fully managed and sustained forest sources. Logging, pulping and manufacturing processes are expected to conform to the environmental regulations of the country of origin.

A catalogue record for this book is available from the British Library.

Library of Congress Cataloging-in-Publication Data
      Effective environmental management in developing countries :
   assessing social capacity development / edited by Shunji Matsuoka.
         p. cm.
      Includes bibliographical references and index.
      ISBN-10: 0–230–54276–X (cloth)
      ISBN-13: 978–0–230–54276–1 (cloth)
      1. Sustainable development – Developing countries. 2. Environmental
   policy – Developing countries. I. Matsuoka, Shunji, 1957–
   HC59.72.E5E34 2007
   333.709172'4—dc22   2007060004

10 9 8 7 6 5 4 3 2 1
16 15 14 13 12 11 10 09 08 07

Printed and bound in Great Britain by
Antony Rowe Ltd, Chippenham and Eastbourne

Contents

List of Tables

List of Figures

List of Photographs

List of Abbreviations

ASEAN	Association of Southeast Asian Nations
BAPEDAL	Environmental Impact Management Agency of Indonesia
CBO	Community-Based Organization
CCCI	Cebu Chamber of Commerce Inc.
CDIAC	Carbon Dioxide Information Analysis Center
CENICA	National Center for Environmental Research and Training of Mexico
CIDA	Canadian International Development Agency
CIESIN	Center for International Earth Science Information Network
DAO	DENR Administrative Order
DDT	Dichloro-diphenyl-trichloroethane
DENR	Department of Environment and Natural Resources of Philippines
DPWH	Public works and infrastructure maintenance department of Philippines
EU	European Union
FDI	Foreign Direct Investment
GBs	Governing boards
HDI	Human Development Index
IEA	International Energy Agency
ISO	International Organization for Standardization
JBIC	Japan Bank for International Cooperation
JICA	Japan International Cooperation Agency
MCWD	Metro Cebu Water District
MDGs	Millennium development goals
MMA	Metro Manila Authority
MMDA	Metropolitan Manila Development Authority
MONRE	Ministry of Natural Resources and Environment
MOSTE	Ministry of Science, Technology and Environment
NAFTA	North American Free Trade Agreement
NGO	Non-Governmental Organization
NPO	Non-Profit Organization
ODA	Official Development Assistance

OECD/DAC	Organisation for Economic Co-operation and Development/Development Assistance Committee
PIDS	Philippine Institute for Development Studies
PROPER	Company performance rating programme
SCA	Social capacity assessment
SCD	Social capacity development
SCEM	Social capacity for environmental management
SEMS	Social environmental management system
Sida	Swedish International Development Cooperation Agency
SWAPs	Sector-wide approaches
UN	United Nations
UNCED	United Nations Conference on Environment and Development
UNCTAD	United Nations Conference on Trade and Development
UNDP	United Nations Development Programme
UNEP	United Nations Environmental Programme
UNESCAP	United Nations Economic and Social Commission for Asia and the Pacific
UNESCO	UN Educational, Scientific and Cultural Organization
UNU	United Nations University
WHO	World Health Organization
WRI	World Resources Institute
WSSD	World Summit for Sustainable Development
WTO	World Trade Organization

Notes on the Contributors

Yumi Fukuhara is Post-doctoral Research Associate at Graduate School for International Development and Cooperation, Hiroshima University. She earned a PhD degree from Kobe University in 2006. Her current research area is urban policies, focusing on sustainability and social capacity development.

Hoetomo is the Deputy Minister for Environmental Compliance in the Ministry of the Environment in the Republic of Indonesia. Before his current appointment, he worked for Ministry of Finance from 1972 to 1997. Since 1997, he has documented various legal drafts including the Law of Natural Resources Management, the Law of Genetic Resources, the Law of Domestic Waste Management, and the Revision of Environmental Management Act.

Senro Imai is a Senior Advisor of the Japan International Cooperation Agency (JICA) and visiting lecturer at Gakushuin University. He is now engaged in a technical cooperation project in Viet Nam, the Philippines, Kenya and China, all of which relate to the capacity development in the field of environmental management.

Eiji Iwasaki is the Team Director of the Environmental Management Team in the Japan International Cooperation Agency (JICA). His team is in charge of technical cooperation projects for capacity development on air pollution control, water pollution control and solids waste management in Central and South America, the Middle East, east Europe and Africa.

Satoru Komatsu is a doctoral student and JSPS researcher at the Graduate School for International Development and Cooperation, Hiroshima University. He has been the research fellow of the Japan Society for the Promotion of Science since August 2006. His research areas include land use modelling, and the economic valuation of environmental and natural resources.

Akifumi Kuchiki is the Executive Vice-President of the Japan External Trade Organization. He is also Guest Faculties Project Professor at Tokyo University and visiting professor at Hiroshima University. He worked at the World Bank as a Senior Economist from 2000 to 2002. He co-edited *Industrial Cluster in Asia*, published by Palgrave Macmillan.

Hebin Lin is a doctoral student at the Graduate School for International Development and Cooperation in Hiroshima University, where she earned a Master's degree in September 2006. Her research interests include the economics of climate change, integrated assessment modelling and sustainable development.

Takashi Matsumura is the Senior Fellow, Environment and Sustainable Development Programme, United Nations University (UNU). He worked at the World Bank as an Environmental Specialist from 1995 to 1998. He earned a Master's degree in science from Tokyo University. His research interests include environment and energy policy integration, sustainable community development, and capacity development for environmental management.

Shunji Matsuoka is Professor in the Graduate School of Asia-Pacific Studies, Waseda University, Japan. He was Professor at the Graduate School for International Development and Cooperation, Hiroshima University, Japan, for many years. He was also Visiting Professor at the University of Malaya, Malaysia, in 1996, and Visiting Researcher at the American University, USA, in 2000. He is currently leading a project of Official Development Assistance (ODA) evaluation funded by the Japan Bank for International Cooperation.

Kazuma Murakami is a doctoral candidate at the Graduate School for International Development and Cooperation, Hiroshima University. He is also an analyst at Mitsubishi UFJ Research and Consulting in Japan. He works for policy design of central and local government. His research areas include environmental policy in Japan, policy evaluation, and capacity development.

Megumi Muto is a Senior Economist, Director, JBIC Institute (JBICI), Japan Bank for International Cooperation (JBIC). She earned her Master's degree in Public and International Affairs from Woodrow Wilson School, Princeton University. Her research interests include effects of physical infrastructure on poverty dynamics, market integration, and human resource outcomes.

Atsushi Ohno is a Post-doctoral Research Associate at Graduate School for International Development and Cooperation, Hiroshima University. He earned a PhD degree from the Department of Economics in Kyoto University in 2006. His research areas include international economics and analysis of international development policy and environmental cooperation.

Metin Senbil is a Post-doctoral Research Associate at the Graduate School for International Development and Cooperation, Hiroshima University. He earned a PhD in transportation planning from Kyoto University in 2003. His research topics cover issues related to sustainable transport and urban development, transportation demand, telecommunications and transportation interactions, quality of life and the environment.

Yoshi Takahashi is an Associate Professor at the Graduate School for International Development and Cooperation, Hiroshima University, where he earned a PhD degree in 2000. His research areas include human resource development in manufacturing sector, industrialization in developing countries, and development assistance.

Katsuya Tanaka is an Assistant Professor at the Graduate School for International Development and Cooperation, Hiroshima University. He earned a PhD degree from the Department of Agricultural and Resource Economics in Oregon State University in 2003. His research areas include land use and watershed modelling, efficiency measurement, and the economic valuation of environments.

Taisuke Watanabe is the Team Director of the Environmental and Social Considerations Review Team, Planning and Coordination Department in Japan International Cooperation Agency (JICA). His speciality is policy development on environmental management, especially pollution control, ranging from theory to practice. Recently, he has begun research work on capacity development in water and solid waste area.

Teppei Yamashita is a Post-doctoral Research Associate at Graduate School for International Development and Cooperation, Hiroshima University. He earned a PhD degree from the Department of Bioresource Sciences in Nihon University in 2007. His current study theme is Capacity Development for Environmental Management in Indonesia.

Preface

Mission

In the 1990s, it became apparent that the replacement approach – the introduction of the knowledge and technology of advanced countries in developing countries in a one-sided manner – had its limitations in the field of international development assistance. The capacity development approach is gradually replacing this conventional approach because it has been highlighted that it is imperative for developing countries to assume ownership for the improvement of their social capacity in order to achieve comprehensive and sustainable development performance. Overall, there has not been any considerable amount of progress made in research and development activities with regard to the capacity assessment technique. However, evidence of some progress can be found in the stakeholder analysis and the institutional analysis carried out by the United Nations Development Program (UNDP), the Canadian International Development Agency (CIDA), and other such agencies.

Our programme, The 21st Century Center of Excellence (COE): The 'COE for Social Capacity Development for Environmental Management and International Cooperation' has developed the social capacity assessment (SCA) methodology with the aim of helping developing countries to achieve sustainable development performance. The SCA methodology involves (1) the definition of the concepts of SCD; (2) the establishment of formal analytical models; and (3) the development of indicators for SCD. In developing countries, the establishment of the SCA methodology will contribute to policy making in terms of effective international development assistance as well as environmental management. Further, developing countries will be able to assess their social capacity using the SCA techniques as well as formulate their own programmes for SCD.

For the practical application of SCA, it is necessary to decide upon a feasible capacity level target and grasp all the capacity development components of the system – external factors such as ODA, the state of the social economy, the social environmental management system (SEMS) and environmental performance. SCA should work as an inexpensive and simple method. It should also be based on scientific research because the developing countries themselves have to apply this method using a participatory approach. Another necessity is the development of self assessment capabilities; this is in order that developing countries themselves can assess their social capacity.

This book provides comprehensive knowledge and techniques of SCA, focusing on environmental management issues faced by developing countries; it also discusses the scientific foundations of the capacity development approach for sustainable development.

Background

The COE programme*

The COE programme was initiated in 2002 by the Japanese Ministry of Education, Culture, Sports, Science and Technology (MEXT) selectively to encourage and support universities in Japan to become world centres of excellence in 11 academic fields through provisions for high quality research and education. This five-year programme is one of the most important Japanese governmental strategies with regard to innovation in universities.

From August 2003, the 'COE for Social Capacity Development for Environmental Management and International Cooperation' at Hiroshima University initiated research on SCD in order to explore new terrain and international environmental cooperation studies. In August 2004, our COE programme – in partnership with the Japan International Cooperation Agency (JICA); the Japan Bank for International Cooperation (JBIC); the Institute of Developing Economies, Japan External Trade Organization (IDE-JETRO); and the National Institute for Environmental Studies, Japan (NIES) – led to the foundation of the Japan Committee on Social Capacity Development for the purpose of researching effective environmental management. This was undertaken because 'bridging research and policy, and/or operation' is indispensable for implementing our research mission.

In order further to develop our theories and strategies, we presented our SCD model and SCA framework at the 4th COE International Symposium in Tokyo in May 2005, the Japan Committee and the World Bank Joint Seminar in Washington D.C. in November 2005, and the 5th COE International Symposium in Tokyo in March 2006. On the basis of these symposiums and seminars, we conducted research and development activities for our SCA methodology and the ways in which it can be used to frame effective aid policies for environmental management in developing countries, particularly Asian countries. This book presents a comprehensive outcome of our works based on the COE program.

* From 2002 to 2004, 274 programs across 11 disciplines in 93 universities were selected for COE. For further information, please visit the JSPS website at http://www.jsps.go.jp. With regard to our COE activities, please view our website at http://home.hiroshima-u.ac.jp/hicec/.

Framework

This book is divided into two parts. The first part provides the theoretical framework of SCA; the second discusses the practical applications of our theory and model.

Part I

In order to analyze social capacity, Chapter 1 discusses the scoping process of capacity development and the entire system. When considering capacity development for environmental management, we must first analyze the characteristics of environmental issues in order to identify the scope of capacity development. Subsequently, we must study the relationship among SEMS (social capacity and institutions), socioeconomic conditions, environmental performance, and external factors such as international aid. This chapter also discusses three important principles of social capacity development; namely, comprehensiveness, ownership and sustainability.

Chapter 2 demonstrates the method by which social capacity development is measured. Benchmarks and indicators (B & I) provide the following: (1) the indicator representing the level of social capacity, and (2) the benchmark providing the threshold of each of the three development stages (the system-making stage, the system-working stage, and the self-management stage). The indicator relies on basic information pertaining to social capacity, which is derived from the results of four analyses – actors–factors analysis, institutional analysis, path analysis and development stages analysis.

Chapter 3 discusses the actors–factors analysis as well as the institutional analysis. The actors analysis reveals the level of social capacity in a given period of time by estimating the capacity of three actors (government, firms and citizens) and the relationships among them. This analysis also reveals substitution and complementary patterns among the three actors and the minimum capacity level required by each actor for their interactions (referred to as 'the critical minimum' or 'benchmark'). The factor analysis focuses on the factors constituting the social capacity – those of P: policies and measures; R: human and organizational resources; and K: knowledge and technology. This provides the capacity level of each of the three factors and their critical minimum. The institutional analysis investigates the institutions, which serve as a basis of social capacity, regulating the three actors. These relevant institutions (referred to as 'the bundle of institutions') imply how current institutions can be changed to promote the development of social capacity. This analysis deals with informal institutions (e.g., social norms) as well as formal institutions (e.g., the legal system) and the interactions between them.

Chapter 4 discusses the development stages analysis and the path analysis. Based on the classification of the three stages of social capacity development – the system-making, system-working, and self-management stages – the development stages analysis reveals the current development stage, the next target, and the approach taken to achieve this target (with the path analysis). Then, it provides us with the most appropriate strategy for development aid. The path analysis reveals the current path for social capacity; subsequently, by setting up the next target, it provides the method by which to achieve this target, including discovering relevant information and conditions. The analysis also investigates the development path among the social capacity level, the socioeconomic background and environmental performance. Based on the relationship and the development path among the social actors, the path analysis provides the capacities of all the actors.

Chapter 5 describes a policy design based on the social capacity development approach. Taking into consideration the result of the SCA, it is possible to design a policy/programme, using which the country can achieve a higher capacity level. This programme should contain the relationship among the actors, the interventions of the factor components (P, R, and K), the timing and the institutional reforms. The scope of the policy/programme described in this chapter includes all three actors: government, citizens and firms. The scope of the project contains not all, but one or two of them. Further, this chapter discusses how to realize the three principles of SCD (comprehensiveness, ownership and sustainability) in effective environmental management.

Part II

Using a flowchart, Chapter 6 describes the construction of a prototype model on how to prioritize policies for poverty reduction and environmental management. There are three stages of economic growth. The first stage involves reducing poverty in order that people can attain a social subsistence level. The second stage entails building social capacity to prepare for sustainable growth. The third stage involves the introduction of growth strategies, including the industrial cluster policy. The author divided the above three stages into the following six steps in order to mitigate the negative impacts on society and the environment: Step 1: attain the social subsistence level; Step 2: attain macroeconomic stability; Step 3: liberalize the economy by structural adjustment programmes; Step 4: develop a capacity that is related to sustainable growth; Step 5: adopt a growth strategy; Step 6: reduce income disparities and improve environmental issues.

In Chapter 7, we discuss how social capacity is viewed as a game that is replayed amongst the government, firms and citizens. Transactions that are repeated over a long period of time between the same actors may lead to a cooperative game, resulting in improved environmental or social infrastructural management. In developing societies, where rules and laws are weak and difficult to enforce, cooperative game solutions are potentially effective strategies to manage issues that involve externalities such as the environment or infrastructure. Addressing SCD in the context of infrastructure management may prove to be more challenging than that for environmental management. In developing societies, the roles of the local and regional governments are more distinct and the civil society tends to be divided into beneficiaries and those that are negatively affected. Case studies conducted in the Philippines have resulted in a policy suggestion: under rapid urbanization, where regional perspective needs to be introduced for infrastructure, improved institutional coordination between the local and regional levels as well as better accountability toward civil society are the keys to success for infrastructure management.

Chapter 8 discusses how a technical cooperation project on SCD in the field of environmental management can be designed. In order to design this project, two capacities can be identified; namely, the current capacity and the capacity to fulfil the requirement by legislations and institutions or the needs of the society (required capacity). Based on the experiences of JICA, the authors have applied SCA to technical cooperation issues. Authors of JICA proposed three steps of capacity assessment for technical assistance in developing countries. The first step is to grasp the general features of the current capacity of the players and the factors influencing the current capacity. The second step is to assess legislations/institutions and players. In this step, special focus is placed on the gap between the current and required capacity. The third step is to set targets of the project on SCD. Following this, Chapter 8 discussed the case of water quality management in the Philippines.

Chapter 9 briefly touches upon SCD for biodiversity conservation issues at the community level. The author analyzes a conservation policy for endangered species from the viewpoint of people-led activities. The Okinawa rail (*Gallirallus okinawae*) is an endangered bird species that is endemic to Yambaru, Okinawa Island, Japan. In 1981, the Okinawa rail was reported as a newly discovered species, and its total population was estimated to be approximately 2,000. A recent survey concluded that its population would soon decrease to less than 1,000. The village of Ada is located at the border of the area inhabited by the Okinawa rail. Approximately 250 people in this village have been working toward coexistence with and the preservation of the Okinawa rail. Innovative activities include establishing community rules for the sound raising of household cats and conducting joint mowing practices. This chapter discusses these community efforts on the SCD model.

In Chapter 10, the author describes several environmental dispute cases in Indonesia, based on extended experiences at the Ministry of Environment (KLH), Indonesia. These environmental cases have several important characteristics: (1) they involve a conflict with development; (2) they commonly involve powerless victims; (3) they require scientific evidence; and (4) to some extent, they escalate into political issues. In 1997, the Environmental Management Law – Law no. 23 – accommodated an instrument of environmental dispute settlement involving the use of alternative dispute resolution (ADR). According to this law, parties have the option of setting their environmental disputes through either an in-court settlement or out-of-court settlement. This instrument has been implemented since the 1980s, as stipulated in the Principle of Environmental Management Law of 1984 – Law no. 4. In 1997, this law was later revised to Law no. 23. This chapter provides examples of a number of environmental dispute cases pertaining to water pollution, which were settled using ADR through the out-of-court mechanism. The chapter also states some of the advantages and limitations of this mechanism. In addition, it puts forth some recommendations in order to gain the optimum use of the out-of court mechanism to settle environmental disputes.

Acknowledgements

As stated above, this book is the output of three years of activity of the 21st Century COE programme: 'COE for Social Capacity Development for Environmental Management and International Cooperation' and the result of a two-year joint study by the Japan Committee on Social Capacity Development. The editor is grateful to JICA, JBIC, IDE-JETRO, NIES and KLH.

The editor is especially thankful to all those who contributed to this volume, beginning with the contributing authors: Akifumi Kuchiki, Megumi Muto, Senro Imai, Taisuke Watanabe, Eiji Iwasaki, Takashi Matsumura and Hoetomo. Many thanks are due to our external evaluation committee, overseas advisory board and the participants invited to our symposiums and seminars: Hidefumi Imura, Yoh Iwasa, Toru Taguchi, Teruyuki Tanabe, Toshihisa Toyoda, Hiromitsu Muta, Akio Morishima, Shoichi Yamashita, Mikimasa Yoshida, Yukio Yoshimura, Keiichi Tango, Keiko Osaki-Tomita, Jun Kukita, Tsutomu Shibata, Yukihiro Shiroishi, Chikako Takase, Sachiko Kuwabara-Yamamoto, Kazumi Goto, Izumi Ohno, Yoshio Okubo, Konrad von Ritter, Kenneth King, Hua Wang, Adriana N. Bianchi and Marian S. delos Angeles.

Finally, thanks are also due to Yumiko Ishikawa, Hiroko Araki, Akemi Kosaka, Noriko Fukuma, Tomoko Sasaki, Sonoko Watanabe, Atsushi Ohno, Yumi Fukuhara and Katsuya Tanaka for their excellent editorial assistance.

SHUNJI MATSUOKA
Ex-Principal Researcher of the 21st Century
Center of Excellence (COE) Programme:
'COE for Social Capacity Development for
Environmental Management and International Cooperation'
Chair of Japan Committee on Social Capacity Development
Professor, Graduate School of Asia-Pacific Studies
Waseda University, Japan

Part I

Social Capacity Development: Theoretical Foundation and Assessment

1
Social Capacity Development for Environmental Management

Shunji Matsuoka and Satoru Komatsu

Overview of capacity development

Issues concerning capacity development

Since the late 1980s, the concept of capacity development has become popular among donor agencies, and, consequently, capacity development in public and private sectors has become the main issue of development (Matsuoka, 2003). As with the aid approach, the replacement approach – that is, the one-sided transfer of knowledge and technology from developed countries to developing ones – was insufficient to deal with the issues regarding international development assistance. Table 1.1 details the problems in the technical cooperation presented by Fukuda-Parr *et al.* (2002) as undermining local capacity, distorting priorities, choosing high-profile activities, fragmenting management, using expensive methods, ignoring local wishes, and fixating on targets.

However, the problems pointed out by Fukuda-Parr *et al.* (2002) are directed at the technical cooperation that is implemented as a replacement approach. Further, they suggested three points for successful technical cooperation: that it should (1) aim at capacity development at a social level, in addition to individual and institutional levels; (2) respect ownership based on the participatory decision-making process; and (3) focus on transforming tacit knowledge to externalized activities through knowledge networks/communities. They revealed that capacity development of developing countries is necessary to improve their social capacity and enable them to achieve sustainable development performance.

Table 1.1   Problems of conventional technical cooperation

Undermining local capacity   Rather than helping to build sustainable institutions and other capabilities, technical cooperation tends to displace or inhibit local alternatives.

Distorting priorities   The funding for technical cooperation generally bypasses normal budgetary processes, escaping the priority-setting disciplines of formal reviews.

Choosing high-profile activities   Donor countries frequently cherry-pick the more visible activities that appeal to their home constituencies, leaving recipient governments to finance the other routine but necessary functions as best as they can.

Fragmenting management   Each donor sends its own package of funds and other resources for individual programmes and demands that recipients follow distinctive procedures, formats and standards for reporting, all of which absorb scarce time and resources.

Using expensive methods   Donor countries often require that developing countries purchase goods and hire experts from them, although it would be far cheaper to source them elsewhere.

Ignoring local wishes   Donor countries pay very little attention to the communities that are supposed to benefit from development activities, local authorities and NGOs, all of whom should comprise the foundation on which to develop a stronger local capacity.

Fixating on targets   Donor countries prefer activities that display clear profiles and tangible outputs. Successful capacity development, on the other hand, is only intrinsically included.

Source: Fukuda-Parr *et al*. (2002).

Capacity development for aid effectiveness

The Paris High-Level Forum meeting was held from 28 February to 2 March 2005 and was attended by many governmental officials and ministers from donor and developing countries. As seen in Table 1.2, the Paris Declaration declared the following five principles that donor and developing (partner) countries should develop: ownership, alignment, harmonization, managing for results and mutual accountability. In this context, especially for the development of alignment, capacity development is a necessary condition for both donor and developing countries to develop country-led strategies. Since ownership is impossible without recipient countries' capacity, capacity development is crucial for achieving other principles; it plays an important role in the development policies.

Table 1.2 Paris Declaration on aid effectiveness

Ownership	Partner countries exercise effective leadership over their development policies and strategies and coordinate development actions.
Alignment	Donors base their overall support on partner countries' national development strategies, institutions and procedures.
Harmonization	The actions of donor countries are more harmonized, transparent and collectively effective.
Managing for results	Managing resources and improving decision-making for results.
Mutual accountability	Donors and partners are accountable for development results.

Source: Adapted from World Bank (2005).

As agreed in the Paris Declaration, capacity development is the responsibility of developing countries, with donor countries playing only a supportive role. In this context, capacity means to plan, manage, implement and account for the results of policies and programmes, all of which are critical for achieving development objectives – from analysis and dialogue to implementation, monitoring and evaluation (World Bank, 2005). Capacity development should not only to be based on a sound technical analysis, but should also be responsive to the broader social, political and economic environments, including the need to strengthen human resources.

However, these explanations do not specify the country that needs to develop its capacity or engage in capacity development. Including the actors' dimensions, these concepts are proposed by many donor organizations such as the Canadian International Development Agency (CIDA), the United Nations Development Programme (UNDP), the Swedish International Development Cooperation Agency (Sida), the Japan International Cooperation Agency (JICA), and so on. To capture the actors' capacity, for example, UNDP (2006) discusses three levels of capacity as individual, organization and enabling environment levels. CIDA defines four levels; individual, organization, sector/network, and enabling environment levels (Bolger, 2000). But these approaches have some limitations in their application to capacity development policy.

Critics of conventional capacity development approaches

These so-called 'individualism approaches' fail to cover the capacity of the entire society because stakeholders targeted under capacity development are, in essence, the social actors. For example, the UNDP approach lacks a middle system that connects organizations with the society, making it difficult to understand the dynamics of micro (individuals and organizations) and macro (society) and the mechanism of adjustment (Matsuoka *et al.*, 2004).

The social actor approach comprising three main actors – government, firm, and citizen as the entire society – has been sufficiently emphasized in this chapter. Table 1.3 presents the definitions of 'capacity' and 'capacity development' as given by CIDA and UNDP in contrast to those

Table 1.3   Definition of capacity, capacity development

Institution	Year	(Social) capacity	(Social) capacity development
CIDA	2000 2004	The ability of individuals, institutions and whole societies to solve problems, make informed choices, order their priorities and plan their futures, as well as implement programmes and projects to sustain them.	The approaches, strategies and methodologies used by developing countries and/or external stakeholders to improve performance at individual, organizational, network/sector or broader system levels.
UNDP	2006	The ability of individuals, institutions and societies to perform functions, solve problems, and set and achieve objectives in a sustainable manner.	Capacity development is (thereby) the process through which the abilities to do so are obtained, strengthened, adapted and maintained over time.
Matsuoka and Komatsu	2006	The ability of social actors to monitor, analyze and evaluate taking problems into consideration, and implement sustainable policy.	The comprehensive, endogenous and sustainable process by which to solve problems in partnership with social actors, interaction with socioeconomic conditions and performance.

Sources: Adapted from Bolger (2000), Lavergne (2004) and UNDP (2006).

Table 1.4   Benefits of capacity assessments

Bring rigour and a systematic method for assessing capacity needs, establishing priorities and sequencing of interventions (as opposed to wishful shopping lists)

More profound systematic challenges, shifting the capacity development question from one of technical cooperation (TC) to a more holistic human development framework

Identify strengths and weaknesses, opportunities for and threats to capacity development

Establish capacity baselines against which to measure, monitor and evaluate progress and performance in capacity development

Make sense of very complex development situations, when it is not always evident where best to intervene to promote capacity development

---

Source: Modified from UNDP (2006).

provided in this chapter. Here, social capacity is defined as the ability of social actors to monitor, analyze, evaluate by taking into consideration various environmental problems and implement sustainable policy. Social capacity development (SCD) is a process in which social actors obtain social capacity; this process is defined as the endogenous and sustainable process that solves problems in partnership with social actors and interacts with socioeconomic conditions and environmental quality.

As discussed above, social actors are responsible for capacities, and capacity development enhances these actors' capacities. Yet, how can we understand the capacity levels or capacity development? Capacity assessment can be a framework to measure the current and desired capacities (the critical minimum level of capacity) and the development path of capacities.

A capacity assessment framework enables the formulation of an effective aid policy. Table 1.4 presents the main benefits expected under good capacity assessment. However, a capacity assessment framework has been disregarded thus far. This point will be discussed under environmental issues with the concept of SCEM.

Social capacity development and social capacity assessment

Capacity development in environment

Discussions on capacity development in environment (CDE) began with an OECD initiative in 1989. In 1992, this concept attracted substantial

Table 1.5　Concept of capacity development in environment

Capacity in environment	Ability of individuals, groups, organizations and institutions, in a given context, to address environmental issues as part of a range of efforts to achieve sustainable development
Capacity development in environment (CDE)	The process by which capacity in environment and appropriate institutional structures are enhanced

Source: Adapted from OECD (1995).

interest at the United Nations Conference on Environment and Development (UNCED, Rio de Janeiro). CDE describes the process by which capacity in environment and appropriate institutional structures is enhanced. Table 1.5 defines capacity in environment and CDE (OECD, 1995). It represents the ability of individuals, groups, organizations, and institutions, in a given context, to address environmental issues as part of a range of efforts to achieve sustainable development (Matsuoka *et al.*, 2004). See Table 1.6 for the main topic of the discussion on CDE. Nevertheless, few significant events concerning these concepts and application for aid policy have occurred in recent times.

However, the discussion on CDE is valuable in bringing up the concept of social capacity for realizing sustainable development; CDE could only cover the capacity in relation to the environment in general as a principle (Matsuoka *et al.*, 2004). The definition or principles of CDE are ambiguous, making it difficult to understand the CDE situation as a development goal.

Social capacity for environmental management

Social Capacity for Environmental Management (SCEM) is a theory that absorbs the CDE concept into a scientific analysis. The SCEM is defined as the ability of social actors to monitor, analyze, evaluate by taking into consideration various environmental problems and implement sustainable environmental policy. The SCD for environmental management is defined as a comprehensive, endogenous and sustainable process to solve environmental problems in partnership with social actors and in interaction with socioeconomic conditions and environmental quality. See Table 1.3 for a comparison of this definition with those of other donors. Figure 1.1 reveals the relationship between social actors in the SCEM structure.

Table 1.6 History of the concept of capacity development in environment

Year	Event	Progress
1989	The Working Party on Development Assistance and Environment (in OECD/DAC)	Beginning of the argument on aid and environment
1992	The United Nations Conference on Environment and Development (UNCED, in Rio de Janeiro)	Institutional building mentioned in Agenda 21
	Taskforce on Capacity Development in Environment (in OECD/DAC)	Established to develop a programme approach of technical cooperation and analytical tools of CDE
1993	International CDE Workshop in Costa Rica	Discussed definition of 'Capacity in Development' and its basic approach
1995	Published Donor Assistance to Capacity Development in Environment (OECD)	Capacity in environment was defined as 'the ability of individuals, groups, organizations and individuals, groups, organizations and institutions in a given setting to address environmental issues as part of a range of efforts to achieve' • Identification of capacity and capability • Improvement of institutional structure • Emphasis on 'process'
1999	Published Donor Support for Institutional Capacity Development in Environment: Lessons Learned (OECD)	The lessons from CDE cooperation • The ambiguous definition of CDE • The importance of CDE in rural areas • Development of the indicator for CDE

Sources: Adapted from Matsuoka and Honda (2002), Matsuoka (2003), OECD (1995) and OECD (1999).

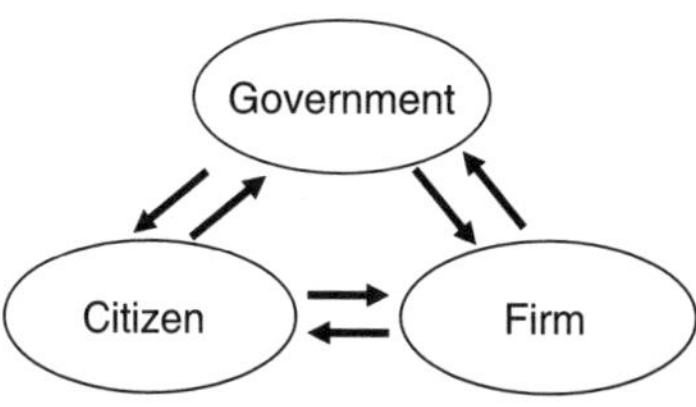

Figure 1.1 Structure of social capacity for environmental management (SCEM)

In addition, SCEM can be developed at both levels: the central government level – which institutes nationwide environmental policies and environmental laws and the local level – which is composed of local governments, firms, and citizens who actually implement the environmental policies and laws (Matsuoka *et al.*, 2004).

Social capacity assessment[1]

As shown in Figure 1.2, the Social Capacity Assessment (SCA) is designed to analyze the interactions in the SCEM and the socioeconomic conditions and environmental quality in a total system. It is believed that the actors will interact in a close relationship as they achieve their capacities, this tendency being identical to the interaction in the SCEM and the socioeconomic conditions and environmental quality. Further, it is believed that the stage shift will occur at the same time as the SCEM development – from the system making stage to the system working stage to the self-management stage. Each stage has a benchmark known as the critical minimum, which means a minimum capacity level necessary for progression to the next stage.

There are three analytical steps in SCA: indicator development, actor–factor analysis and development stage analysis. Indicator development enables us to understand the current position and situation of environmental management capacity in each country. The actor–factor analysis enables a detailed analysis of indicator development in order to determine the strength or weakness of an actor or factor with regard

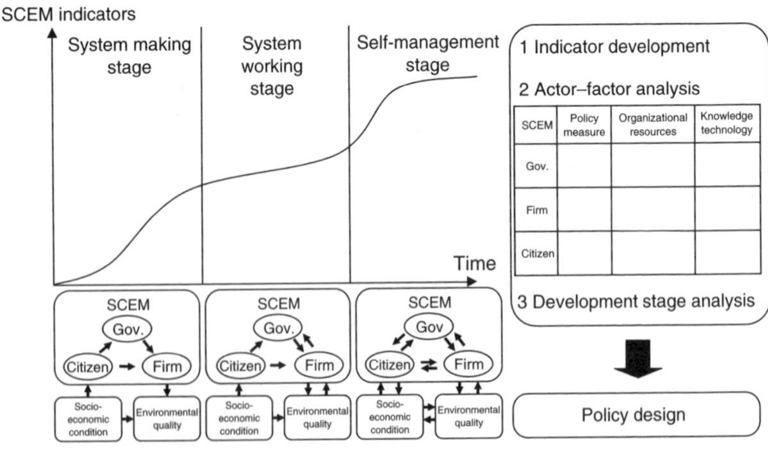

Figure 1.2   SCEM and social capacity assessment

to capacity development. The development stage analysis aims first at specifying the development stage, based on benchmarks, and then presenting the development process and direction for further development. The analysis based on each of these steps is provided from Chapters 2 to 4.

Principles of social capacity development

The principles of SCD comprise comprehensiveness, ownership and sustainability.

Comprehensiveness

The use of the term 'comprehensiveness' means that all elements of development – social, structural, human, governmental, environmental, economic and financial – are included for development and poverty reduction (World Bank, 2006). This meaning is applied to the Comprehensive Development Framework (CDF), which is an approach to guide development and poverty reduction, including the provision of external assistance applied by World Bank (World Bank, 2006).

The approach in this chapter is far removed from the CDF approach, but it will consider the comprehensiveness approach in the SCD process. SCD includes three social actors and factors. This approach establishes a new sub-system known as the social environmental management system (SEMS), which is a social system that operates the SCEM and is affected by the relevant socioeconomic conditions and environmental quality.

Thus, 'comprehensiveness' embraces a different approach from those adopted by the World Bank and SCD. 'Comprehensiveness' in the World Bank approach relates to every resource that contributes to development. 'Comprehensiveness' in SCD formulates a social system for analysis. This comprehensiveness does not simply sum up the social actors and factors. Instead, it creates synergy effects and social sub-systems using the SCD approach.[2]

Ownership[3]

Where ownership exists, social actors are motivated by the policy-making and project implementation process, with the support of donor countries that align their strategies in accordance with recipient countries' needs. According to Jones and Williams (2002), national ownership refers to the level of commitment to development policies and programmes existing within a country. National ownership exists when (1) development policies and programmes are defined and directed by

Table 1.7   Conventional explanation of national ownership

National ownership refers to the level of commitment to development policies and programmes existing within a country. Country ownership can be said to exist where the following conditions are present:

Development policies and programmes are defined and directed by the recipient country government;

Donors do not impose their ideas but respond to the initiatives taken by the government;

Commitment to a country's development policies and programmes is widely shared across society.

Source: Jones and Williams (2002).

the recipient country government; (2) donor countries do not impose their ideas but respond to the initiatives taken by the government; and (3) commitment to a country's development policies and programmes is widely shared across society (Jones and Williams, 2002; see Table 1.7 for a detailed explanation). According to CIDA (2000), local ownership implies that development strategies must be conceived by recipient countries – their governments and their people – and they must reflect their priorities, rather than those of donor countries. The term 'local' in 'local ownership' embraces social actors who engage in the development process. However, this idea is donor-driven, which leads to the following problem: the concept of ownership is applied for capacity development.

When considering capacity development in developing countries, these countries may be evaluated as having a low capacity because they do not have sufficient ownership. Sufficient ownership can help in the planning of development policy and its subsequent implementation, and can be regarded as a high capacity. If donor countries demand sufficient ownership as a condition for offering assistance with capacity development, they would tend to assist countries that have a relatively high capacity for capacity development. Hence, developing countries regarded as having a low capacity or low ownership would be unable to receive assistance and would continue to be in a low capacity situation. This assumption leads to a contradiction.

Moreover, the SCD process ensures the ownership of developing countries by encouraging the monitoring, analysis, problem-specific evaluation and implementation of sustainable policy regarding the self-learning process of the social actors. Donor countries help developing countries with this self-learning process. Further, social actors are motivated to adopt the participatory approach not only for its own sake but also in order to receive assistance from donor countries. This contributes to long-term ownership in developing countries.

Developing countries are by nature diversified societies. Local owner-ship and involvement by donor countries should take this factor into consideration. Consequently, the ownership promised by SCD will lead to development.

Sustainability

Sustainability[4] is an important concept with regard to long-term capacity development. Sustainable development, which reveals the sustainability concept in the development field, is defined as a development that satisfies the needs of the present without compromising the ability of future generations to fulfil their own needs (World Commission on Environment and Development, 1987).

Sustainability comprises three dimensions: environmental, economic and social (environmental – focusing on the resources of the natural physical environment; economic – paying attention to the growth of economies and their structures for humans; and social – embracing social development aspects for human capability or human rights). See Table 1.8 for a detailed explanation.

The UNCED proclaims that the issue of sustainable development in developing countries and in international actions should be addressed as a special priority (United Nations, 1992). One of the Millennium Development Goals (MDG), the development goals adopted in the 2000 General Assembly, is to ensure environmental sustainability. This shows that sustainable development or sustainability is given high priority in international development dialogue.

Further, the SCD process tries to ensure sustainable development. It aims to receive the long-term benefit of capacity development in order to develop overall social capacity. The total system comprises the SCEM, the socioeconomic conditions and environmental quality and attempts development according to the preconditions of a country, its histori-cal background and other structural conditions. This process pursues the development of a sustainable future for intergenerational equity throughout the country. This process, which engages all stakeholders, can be a huge contributor to sustainable development.

Scoping of social capacity and the total system analysis

Scoping of environmental issues

Scoping of environmental issues means the categorization of environ-mental issues into the following three dimensions: characteristics, time

Table 1.8   Three dimensions of sustainable development

---

1   Environmental dimension
The economic processes of production and consumption were provided by resources from the natural physical environment. These resources are

Natural resources: Conventional type includes non-renewables such as minerals, renewables such as forests, and all forms of energy.

Environmental resources: Services can be divided into contributions for 'immediate human consumption' and their 'consumption processes'. The former services sustain the biological basis of human life and well-being as well as provide for the enjoyment of natural resources by people. The latter services are derived mainly from the absorptive capacities of the physical environment and as such contribute to human well-being.

2   Economic dimension
The growth of economies and their structural transformations have always been recognized as being at the core of development. They still are the most important preconditions for the fulfilment of human needs and for any lasting improvements in living conditions.

3   Social dimension
Seen from a broad angle, development encompasses the strengthening of the material income base as well as the enhancement of capabilities and the enlargement of choices. Such a view of development clearly transcends the narrow 'development as economic growth' concept and also emphasises the importance of social development in the context of sustainable development.

Source: Adapted from UNIDO website.

and space (Table 1.9). This categorization is conducted before proceeding to the SCA stage, which focuses on environmental issues. This enables the grasping of environmental issues in a concrete manner. For example, under 'time' air, water and noise pollution, in addition to municipal waste, are categorized as flow-type pollution, which will disappear in the short term. Hazardous waste, forest, ecology and landscape are categorized as stock-type pollution, which will sustain and will require long-term strategies for pollution control.

As shown in Figure 1.3, one dimension can be added by characterizing the following issues with economic growth: poverty, industrial pollution and consumption (World Bank, 1992; Bai and Imura, 2000). The poverty-related issue improves with economic growth, which enhances the investment in a living infrastructure (Matsuoka, 2003). Industrial pollution-related issues consist of mostly air and water pollution, which do not improve when countries deliberately introduce policies

Table 1.9 Scoping the environmental issues: example

Characteristics	
Green issues	Soil, flora and fauna, ecology, landscape, forest, biodiversity
Brown issues	Air pollution, noise, vibration, bad odours, waste
Blue issues	Water quality (inc. underground water)
Time	
Flow	Air, water, noise, municipal waste
Stock	Hazardous waste, forest, ecology, landscape
Space	
Global	Global warming, acid rain
National	Air pollution
Province	Air pollution, industrial waste, river basin management, forest
City	Municipal waste, air pollution, land use, urban planning

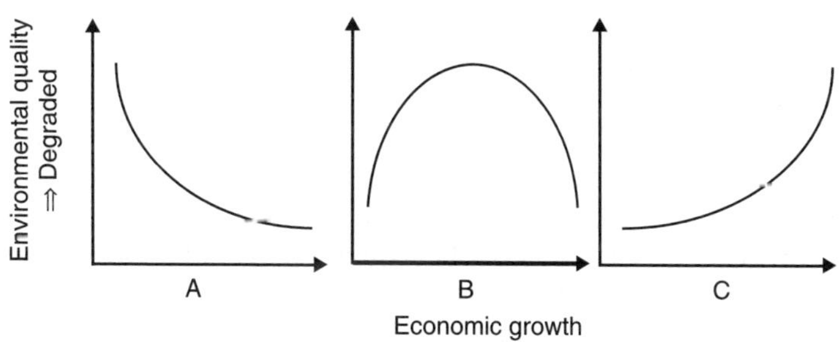

Figure 1.3 Economic growth and environmental issues

Note. A Poverty-related issues: access to safe drinking water and sanitation
B Industrial pollution-related issues: SO_X, PM_{10}
C Consumption-related issues: municipal waste, CO_2

Source: Bai and Imura (2000).

for dealing with environmental problems (World Bank, 1992). The shape of the inverted U-curve is known as the Environmental Kuznets Curve. Consumption-related issues include municipal waste, CO_2 and so on. 'In these cases, the abatement cost is relatively expensive and the cost-associated emissions and wastes are not yet perceived as high – often because they are borne by someone else' (World Bank, 1992).

Capacity assessment and capacity development can be made easier with this categorization. In contrast to developed countries, where environmental issues have shifted from poverty and industrial pollution to consumption, developing countries affect all environmental issues simultaneously. For example, in Jakarta, Indonesia, since air pollution, water pollution, municipal waste and forest logging are simultaneously affected, they become serious concerns. Further, flow-type pollution can also be stock-type pollution. In many developing countries, issues concerning water pollution and municipal waste are dealt with in the absence of proper disposal methods, which leads to the emission of harmful substances over a long period of time. The presence of these issues in developing countries should be taken into account during the scoping process.

Identifying stakeholders: scoping of actors

Scoping of actors helps to identify the stakeholders involved in each environmental issue. This categorization is conducted before performing the SCA, which focuses on stakeholders. For example, if provincial local governments are the most appropriate social actors for solving particular issues, then the role of the central government or city government is comparatively less important. For SO_X reduction, the national government conducts nationwide regulation and the provincial government conducts region-specific regulation, because SO_X concentration is highly affected by its locations. The incentive-based reduction of emission is achieved through the activities carried out by firms, and the capacity of citizens is constituted by their careful observations – monitoring every activity like a watchdog.

This step clarifies the level of commitment to environmental issues. See Table 1.10 for an example of the actor scoping method for air pollution. Note than even if the issues are clarified in a single governmental organization, coordination between other organizational levels continues to be a concern.

Determining how the assessment is performed

The combined use of the above steps enables us to understand capacity in a more concrete manner. Environmental issues are captured through environmental scoping, and the stakeholders involved are characterized through the actor scoping method. The latter is a capacity scoping method for focusing on issues that need to be solved.

Total system analysis

The total system analysis is designed to analyze the interactions between SCEM and the socioeconomic conditions and environmental quality.

Table 1.10   Identify stakeholders: example of air pollution

Social actors	Stakeholders	Commitment
Government	1 Central government (national) Ministry of the environment, environmental divisions	W
	2 Local government (province) Environmental management sectors	W
	3 Local government (cities) Environmental management sectors	S
Firm	1 Firms (emit industrial pollutions), service sector, agriculture	S
	2 Firms (environmental services), consulting, research,	S
	3 Industrial unions	W
Citizen	1 Citizens	S
	2 CSO (Civil Society Organization), NGO (Non-Governmental Organization), NPO (Non-Profit Organization), CBO (Community-Based Organization)	S
	3 Researchers, mass media	W

Note: W represents weak and S represents strong commitment to management.

The SCEM of a country is constrained by the existing socioeconomic conditions and environmental quality; therefore, it is necessary to have a total framework representing the society. This relationship is presented in Figure 1.4.

Honda *et al.* (2004) analyzed the relationship among these three components in 47 Japanese prefectures. From among these analyses related to several environmental issues, we consider the cases of SO_2, NO_2 and SPM. The analysis is carried out using the Granger Causality Test and is based on the data collected for the period ranging from 1982 to 2000. Figure 1.5 confirms the existence of interrelationships between the three components for 23, 20, and 24 prefectures out of 47 in the cases of SO_2, NO_2 and SPM respectively.

'External factors' are the factors outside a country such as aid, external trade or international treaty. These factors are dependent not only on developing countries, but also on donor countries, developed countries and international organizations. To follow the steps recommended in the SCA design, it is important to consider these external factors. For example, OECD (2001) inferred that trade, aid and finance

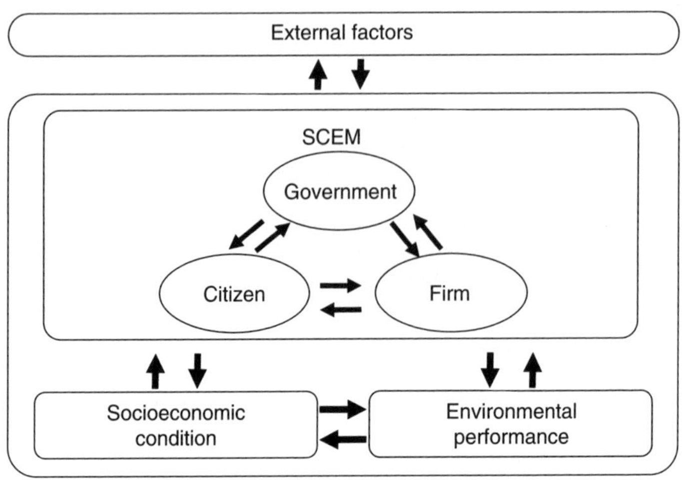

Figure 1.4   Total system of SCEM

	A	B	C	D	Others	Total
	SCEM ECON⇄ENV	SCEM ECON→ENV	SCEM ECON←ENV	SCEM ECON⇄ENV		
SO_2	23	1	11	3	9	47
NO_2	20	2	12	2	11	47
SPM	24	1	2	2	18	47

Figure 1.5   The structure of SCEM: the case of air pollution in Japan

Note: Numbers in the table indicate the numbers of prefectures (totalling 47).

Source: Honda *et al.* (2004).

communities are developing more coherent strategies to help developing countries integrate with the global economy. This approach is known as a 'trade capacity development', which enhances the ability of developing countries to manage trade. (See Table 1.11 for a detailed explanation on trade capacity development.)

Table 1.11   Trade capacity development

Collaborate in formulating and implementing a trade development strategy that is embedded in a broader national development strategy

Strengthen trade policy and institutions – as the basis for reforming import regimes, increasing the volume and value-added of exports, diversifying export products and markets and increasing foreign investment to generate jobs and exports

Participate in, and benefit from, the institutions, negotiations and processes that shape the national trade policy and the rules and practices of international commerce

---

Note: OECD/DAC use 'capacity building' instead of 'capacity development'.

Source: Adapted from OECD/DAC (2001).

Social capacity assessment and policy design

Social capacity assessment: methodological steps

Indicator development

Indicators are developed for conveying current information regarding the SCEM. Indicator development describes the efficiency-based SCEM indicators in ten Asian countries. Detailed methodology and empirical applications will be provided in Chapter 2.

Actor–factor analysis

By analyzing the capacity of social actors and their interrelations, the actor analysis evaluates social capacity at a given time. The factor analysis, on the other hand, focuses on the factors of social capacity; that is, policies and measures, human and organizational resources, and knowledge and technology. The actor–factor analysis enables a detailed analysis of indicator development to determine the actor or factor that is strong or weak in relation to capacity development. Based on the actor and factor analyses, the actor–factor matrix is formulated (please refer to Table 1.12). The methodology applied in the actor–factor analysis is given in Chapter 3.

Development stage analysis

The development stage analysis first aims at specifying the development stage based on benchmarks and then presenting the development process and the direction for further development. The three stages of development are system making, the system working and self-management.

Table 1.12   Actor–factor matrix

Factors\Actors	Policy and measures (P)	Organizational resources (R)	Knowledge and technology (K)
Government (G)	G_P	G_R	G_K
Firm (F)	F_P	F_R	F_K
Citizen (C)	C_P	C_R	C_K

The system making stage is the fundamental basis of the SCEM where governmental institutions and environmental policy are developed. The system working stage is where the government and firm sectors gain strength by setting up incentives for pollution abatement, following which levels of industrial pollution improve after reaching their peak (Matsuoka, 2003). The self-management stage is one that requires a comprehensive environmental policy since it leads to the emergence of consumption related issues. This stage requires harmonious relationships between governments, firms and citizens in order to ensure efficient environmental management (Matsuoka, 2003). This analysis also provides information regarding the 'critical minimum'; that is, the capacity level requiring each social actor satisfactorily to perform the social system functions. The analytical results highlight certain preconditions that clarify the appropriate quantity, quality and timing of input in order to enable development and aid policies to be implemented as programmes. (For details, see Chapter 4.)

Policy design based on social capacity assessment

This section describes the policy design for social capacity development. Based on the SCA framework, we develop the aid approach needed to identify the target level of capacity and to provide specific strategies to achieve that target. The policy presents an overall package that comprises: (1) the relationship between the social actors; and (2) the input resources – their quantity and timing. (For details, see Chapter 5.)

This policy design differs from conventional stand-alone projects in many aspects. This approach can ensure: (1) comprehensiveness; (2) ownership; and (3) sustainability. With regard to comprehensiveness, this approach includes all social actors and their capacities can be developed into social systems under the relevant socioeconomic conditions and environmental quality. Concerning ownership, the social actors are motivated by the increase in their rights on policy commitment, with assistance from donor countries that align their strategies with recipient countries' needs. Regarding sustainability, policies developed under the SCA can embrace long-term implementation in order to achieve the global social system for sustainable environmental development. Table 1.13 shows a detailed comparison between

Table 1.13   Principles of aid effectiveness on stand-alone projects and SCA-based policy

Principle	Stand-alone projects	SCA-based policy
Local ownership	Projects are often supply led. Even when they are demand led, they may be undertaken in response to demands from a single local partner.	SCA-based policy is based on the development countries' ownership to which all stakeholders are committed.
Donor coordination	Limited donor collaboration leads to high transaction costs, possible duplication of efforts and sub-optional identification of priorities.	Donors can involve the projects under the capacity development initiative to achieve sustainable development performance. Under the recipient countries' leadership, the project duplication may be reduced.
Partnership	Projects are often managed directly by executing agencies or project implementation units that maintain control over the project and are held accountable for results.	Donor and developing countries can harmonize their policy alignments to achieve the same developmental goals. Donor countries will prepare the corresponding projects if the developing countries cannot fill the capacity gap by themselves.
Attention to institutional development, governance issues and civil society participation	Projects attempt to ensure success by establishing project-specific control mechanisms. Thus, they attempt to bypass, rather than solve, certain institutional weaknesses.	As SCA-based policy embraces all stakeholders committed to the development procedure, social actors (i.e., government, firm, citizen) can take initiatives and be motivated by active participation. Institution and local governance issues are necessary conditions for development goals.

Continued

Table 1.13 Continued

Principle	Stand-alone projects	SCA-based policy
Result-based approach	Attention is focused on the success of the projects themselves, even though other conditions necessary to the achievement of development results may not be met. One thus faces the possibility that development cooperation may be successful at the project level, while failing to promote development more generally.	SCA-based policy aims at filling all capacity gaps in the recipient country. We can evaluate each project as the contribution for social capacity development rather than the success of individual projects.

Source: Lavergne and Alba (2003).

stand-alone projects and the SCA-based policy approach for aid effectiveness. Table 1.13 also shows a detailed comparison based on (1) local ownership; (2) donor coordination; (3) partnership; (4) attention to institutional development, governance issues and civil society participation; and (5) a result-based approach.

In recent times, budget support or sector-wide approaches (SWAPs) for social capacity development can be classified as one of the approaches adopted by the programme. Budget support is an aid modality that is provided in support of a governmental programme (DFID, 2002). The sector-wide approach is defined as an effort to channel the assistance received from donor countries to a sector within a common management and planning framework so that the agreed sector strategy can be implemented (Jones and Lawson, 2000). (See Table 1.14 for the definitions and principles of each modality.) This characterization, however, can be a tool for achieving social capacity development; these approaches do not consider the involvement of social actors. For example, Jones and Lawson (2000) discuss that, in the application of sector-wide approaches, one reaction to a weak governmental capacity has been to increase direct donor funding to NGOs that provide services complementing or substituting governmental services. The change of aid partner is a common occurrence; however, if capacity assessment is not implemented,

Table 1.14   Aid modalities under SCA

	Definition	Elements
Budget support	General budget support – a contribution to the overall budget and any conditionality focuses on policy measures related to overall budget priorities. Sector budget support – a contribution notionally earmarked to a sector/sectors and any conditionality relates to these sectors.	1 Funds are provided in support of a government programme that focuses on growth and poverty reduction, and transforming institutions, especially budgetary. 2 Funds are provided to a partner country to spend using its own financial management and accountability systems.
Sector-wide approaches (SWAP)	An effort to bring donor support to a sector within a common management and planning framework for implementing an agreed sector strategy.	1 The sector-wide approach is a process, not a particular instrument or programme. 2 At the core of the sector-wide approach are the sector strategy and the public expenditure programme that supports it. 3 A common management and planning framework may be involved, but it does not necessarily imply a common pool funding mechanism.

Sources: DFID (2002) and Jones and Lawson (2000).

then the demarcation and service transfer between governments and NGOs cannot be discussed. Currently, the SCA-based policy design is given high priority, followed by the sector-wide approach and other modalities.

Conclusion

This chapter defined the new concept of SCEM formed by three social actors – government, firm and citizen – and their interrelationships. The SCEM can be discussed under the capacity development approach and sustainable development. Measuring the SCEM between the socio-economic conditions and environmental quality can be a capacity assessment framework for developing an effective aid approach.

The analysis techniques will be discussed in Chapters 2, 3 and 4, dividing these techniques into three analyses. The policy formulation methods applicable to the SCA methods will be discussed in Chapter 5.

Notes

1 Although there are many capacity assessment tools provided by the UNDP, all of them are based at the individual level, the organizational level, and the enabling environment level. (See UNDP (1998) and UNDP (2006).)
2 If the Social Capacity Development approach contributes to the overcoming of the 'macro–micro paradox', which implies that aid works at the micro (project) level but not at the macro (policy) level, then aid cannot contribute to sustainable development at the country level. (See Bigsten *et al.* (2005) for further discussion.)
3 Lavergne (2003) explains local ownership as a bundle of rights: the right to set an agenda, the right to allocate resources (including external resources) and the right to design and implement development programmes (policies). When we consider ownership as a bundle of rights, it will gradually be established as the developing countries gain rights.
4 CIDA provides the development policy for environmental sustainability, which realized this concept in the development field. (See further details in CIDA (1992).)

References

Bai, X. and H. Imura (2000) 'A Comparative Study of Urban Environment in East Asia: Stage Model of Urban Environmental Evolution', *International Review for Environmental Strategies*, 1, 1: 135–58.

Bigsten, A., J.W. Gunning, and F. Tarp (2005) 'The Effectiveness of Total ODA: An Evaluation Proposal', Paper presented at OECD/DAC Department Evaluation Network, Evaluating Total ODA Impact, Stockholm, 10–11 November 2005.

Bolger, J. (2000) 'Capacity Development: Why, What and How?', *Capacity Development Occasional Series*, 1, CIDA: 8, http://remote4.acdi-cida.gc.ca/cd (accessed on 31 August 2006).

CIDA (Canadian International Development Agency) (1992) 'CIDA's policy for Environmental Sustainability', Canadian International Development Agency: 15.

CIDA (Canadian International Development Agency) (2000) 'Canada Making a Difference in the World: A Policy Statement on Strengthening Aid Effectiveness', Canadian International Development Agency.

DFID (Department for International Development) (2002) DFID Workshop Handbook on Direct Budget Support and SWAps, in A. Alba and Lavergne, R. (2003) *LENPA Glossary of Frequently-Used Terms Under Program-Based Approaches*,

13 November 2003, CIDA: 35, http://remote4.acdi-cida.gc.ca/cd (accessed on 21 August 2006).

Fukuda-Parr, S., C. Lopez, and K. Malik (eds) (2002) *Capacity for Development – New Solutions to Old Problems* (London and Sterling, VA: Earthcan, UNDP): X and 284.

Honda, N., S. Matsuoka, and K. Tanaka (2004) 'Causal and Structural Analysis on Social Capacity Development for Environmental Management in Japan's Air Pollution Problems', *Discussion Paper Series* (COE for Social Capacity Development for Environmental Management and International Cooperation, Hiroshima University), 2004–7: 35 (in Japanese).

Jones, S. and A. Lawson (2000) 'Moving from Projects to Programmatic Aid', *OED Working Paper Series*, 5: 35.

Jones, S. and G. Williams (2002) 'A Common Language for Managing Official Development Assistance: A Glossary of ODA Terms': 24, http://www.opml.co.uk/publications/development_policy/a_common_languag.html (accessed on 25 August 2006).

Lavergne, R. (2003) 'Local Ownership and Changing Relationships in Developing Cooperation': 9, http://www.ccic.ca/e/docs/002_aid_2003-03_local_ownership_and_changing_relationships.pdf (accessed on 31 August 2006).

Lavergne, R. (2004) 'Capacity Development in CIDA's Bilateral Programming: A Stocktaking – Research Results', 19 January 2004: 41, http://remote4.acdi-cida.gc.ca/cd (accessed on 21 August 2006).

Lavergne, R. and A. Alba (2003) 'CIDA Primer on Program-Based Approaches, Canadian International Development Agency': 63, http://www.acdi-cida.gc.ca/INET/IMAGES.NSF/vLUImages/CapacityDevelopment2/$file/Program%20Based%20Approaches-E.pdf (accessed on 22 August 2006)

Matsuoka, S. (2003) 'Social Capacity Development and Environmental Management', Matsuoka, S. and A. Kuchiki (eds) (2003), *Social Capacity Development for Environmental Management in Asia: Japan's Environmental Cooperation after Johannesburg Summit 2002* (Chiba, Japan: Institute of Developing Economies).

Matsuoka, S. and N. Honda (2002) 'Environmental Cooperation and Capacity Development: Review of the Concept of Capacity Development in Environment (CDE)', *Journal of International Development Studies*, 11, 2: 149–72 (in Japanese).

Matsuoka, S., S. Okada, K. Kido and N. Honda (2004) 'Development of Social Capacity for Environmental Management and Institutional Change', Proceedings of the Second International Symposium of the Social Capacity Development for Environmental Management and International Cooperation in Developing Countries (13 January 2004), Hiroshima University, Hiroshima, Japan: 1–23.

OECD (Organisation for Economic Co-operation and Development) (1995) 'Donor Assistance to Capacity Development in Environment', *Development Co-operation Guidelines Series*, OECD: 12, http://www.oecd.org/dataoecd/33/55/1919786.pdf (accessed on 21 August 2006).

OECD (Organisation for Economic Co-operation and Development) (1999) 'Donor Support for Institutional Capacity Development in Environment: Lessons Learned, OECD': 35, http://www.oecd.org/dataoecd/10/27/2667310.pdf (accessed on 31 August 2006).

OECD (Organisation for Economic Co-operation and Development) (2001) *The DAC Guidelines: Strengthening Trade Capacity for Development*, OECD, p. 68, http://www.oecd.org/dataoecd/46/60/2672878.pdf (accessed on 31 August 2006).

UN (United Nations) (1992) Report of the United Nations Conference on Environment and Development, *Rio de Janeiro*, 3–14 June 1992.

UNDP (United Nations Development Programme) (1998) 'Capacity Assessment and Development in a Systems and Strategic Management Context', Technical Advisory Paper 3, UNDP/BDP/Management Development and Governance Division: 21.

UNDP (United Nations Development Programme) (2006) *Capacity Assessment: Practice Note*: 35, http://www.capacity.undp.org/index.cfm?module=Library&page=Document&DocumentID=5510 (accessed on 21 August 2006).

World Bank (1992) 'World Development Report: Development and the Environment, Volume 1' (New York: Oxford University Press): 322 http://www.wds.worldbank.org/external/default/main?pagePK=64193027&piPK=64187937&theSitePK=523679&menuPK=64187510&searchMenuPK=64187511&siteName=WDS&entityID=000178830 9810191106175 (accessed on 22 August 2006).

World Bank (2005) 'Paris Declaration on Aid Effectiveness: Ownership, Harmonisation, Alignment, Results and Mutual Accountability': 12, http://www1.worldbank.org/harmonization/Paris/FINALPARISDECLARATION.pdf (accessed on 21 August 2006).

World Bank (2006) 'World Bank Strategies on Comprehensive Development Framework', http://www.worldbank.org/cdf/ (accessed 25 August 2006).

World Commission on Environment and Development (1987) *Our Common Future* (Oxford: Oxford University Press): 400.

2

Indicator Development of Social Capacity for Environmental Management

Katsuya Tanaka and Hebin Lin

Why should we develop an indicator?

Introduction

As described in Chapter 1, the Social Capacity for Environmental Management (SCEM) is the capacity to manage and solve environmental problems through an endogenous process involving three actors (governments, firms and citizens) and their interactions (Matsuoka and Kuchiki, 2003). Many traditional arguments deal with causal relations between environmental performance (e.g., SO_2 emission) and socioeconomic conditions (e.g., GDP per capita). The SCEM approach extends such arguments by integrating the SCEM into a traditional framework. Thus, environmental performance is determined by socioeconomic conditions, SCEM and their interactions. Figure 2.1 illustrates this relationship.

The capacity development approach has gained greater attention from various environmental perspectives since the 1990s. However, many of the existing relevant studies are limited in terms of either theory or empirics. Some studies are highly conceptual and do not demonstrate the empirical application of those concepts using the existing data. Some other studies present an empirical measurement of capacity development by developing several indicators. However, many such studies provide little relationship between such indicators and environmental performance. In order to obtain a valid measurement of capacity, more statistically sound and reliable indicators need to be developed.

This section proposes another approach toward the elicitation of the causal relationship between environmental performance, socioeconomic conditions and SCEM capacity indicator development. Our approach is different from existing studies in several respects. First,

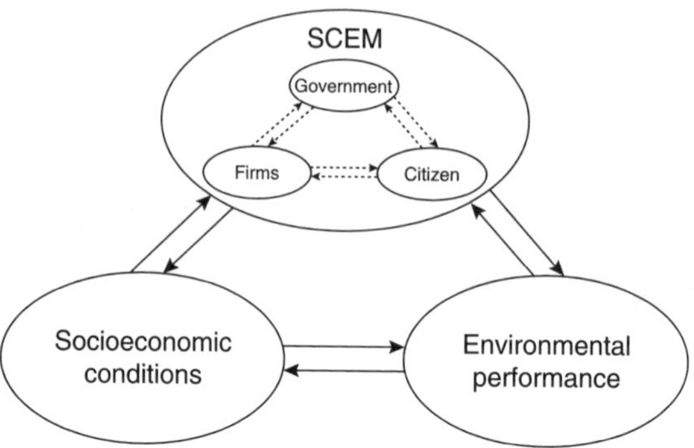

Figure 2.1   SCEM under the total system

environmental performance is measured in terms of efficiency, not in terms of a direct measure of pollutant emission. Second, a statistical investigation is conducted on the relationship between three major components – environmental performance, socioeconomic conditions and SCEM. Third, the SCEM indicator is developed using relevant variables and their statistical significance and elasticities. Thus, the indicator we present in this chapter is statistically valid compared with existing indicators.

Existing indicators

Before presenting the SCEM indicator framework, we summarize and list several major indicators dealing with environmental and natural resources. Overall, there is an extensive number of studies to quantify the capacity of environmental performance and sustainability. Many of these studies either select the relevant indicator variable and evaluate its progress (the selection approach) or aggregate the scores obtained from checklists or variables related to the environment of sustainability (the aggregation approach). We introduce MDG-7 (the Target 9 indicator) as the selection approach. In addition, we introduce air quality management and assessment capabilities and the environmental sustainability index (ESI) as aggregation approaches.

MDG-7, Target 9 indicators

One of the most important innovations of the MDG approach is its ability to make governments more accountable for their performance in improving human well-being. By defining goals and measuring progress in clear, straightforward language, the MDG makes it easy for civil society groups to evaluate the progress towards human development goals and to issue a public 'report card' on a government's success or failure.

Unfortunately, the lack of clear, comprehensive targets and indicators for measuring the capacity of ecosystems to provide a sustainable environmental income for the poor means that the 'accountability effect' of the MDG approach is not yet applicable to the world's environmental goals. Until the environmental framework of the MDG is mended, the short-run progress towards the other goals is at risk of being unsustainable. For this reason, it is recommended that MDG-7 (Target 9) be updated.

Air quality management and assessment capabilities

The United Nations Environmental Programme (UNEP) and the World Health Organization (WHO) propose the aggregation approach of environmental management capacity measurement in their report *Air Quality Management and Assessment Capabilities in 20 Major Cities* (UNEP/WHO, 1996). Their approach measures environmental management capacity in 20 major cities by assessing the number of indicators obtained from four primary categories: (1) Air quality measurement capacity; (2) data assessment and availability; (3) emissions estimate; and (4) management enabling capabilities. They prepare a checklist for each category and request each city to provide the current position regarding air quality monitoring. The sum of the points of checked items under each category is set at 25 points. Thus, this aggregation approach evaluates the capacity of air quality management from the aggregated score of indicators, which takes a value between 0 and 100.

Environmental Sustainability Index

The ESI benchmarks the ability of nations to protect the environment over the next several decades. In addition, in terms of sustainability, the ESI provides the following: (1) a powerful tool for situating environmental decision-making on a firmer analytical footing; (2) an alternative to GDP and the Human Development Index (HDI) for gauging the progress of a country; and (3) a useful mechanism for benchmarking environmental performance (Esty *et al.*, 2005). In order to achieve this goal, the ESI is constructed by integrating 76 data sets – tracking natural resource endowments, past and present pollution levels, environmental management efforts and the capacity of a society to improve its environmental performance – into 21 indicators of environmental sustainability. The indicators and variables on which they are constructed build on the well-established Pressure-State-Response (PSR) environmental policy model.

These indicators permit comparison across a range of issues, which fall into the following five broad categories: (1) environmental systems; (2) reducing environmental stresses; (3) reducing human vulnerability to environmental stresses; (4) societal and institutional capacity to respond to environmental challenges; and (5) global stewardship.

Overall, the ESI, with its emphasis on relative rankings, provides a mechanism for establishing context and for understanding what is possible in terms of policy progress. Indeed, it turns out that comparisons to relevant peer countries are particularly important in goal setting, identifying best practices in both policy-making and technology adoption, and spurring competitive pressure for improved performance.

Outline

The next section introduces the manner in which the SCEM indicator is developed. Each of the three steps involved in the development of the SCEM indicator is described in detail. We then present two empirical applications. These applications reveal how our framework (see pp. 8–11) is used empirically with the existing data. The first application estimates the SCEM indicator and compares its progress among eight Asian countries using international panel data. The second application presents country-specific analysis using province-level statistical data in China. The last section summarizes this chapter and draws conclusions from prior sections. We also present some limitations of our approach and the manner in which these are resolved using alternative approaches.

Three steps involved in indicator development

This study develops the SCEM indicator using the three-step modelling approach. The first step estimates environmental performance by measuring efficiency. We present two different efficiency measures in this section. The first measure is environmental efficiency (EE), which focuses only on the emission of pollutants. The second is balanced growth efficiency (BGE), which estimates the efficiency of sustainable economic development by dealing with both output and pollutant emission. The estimated environmental performance is then used in the second step to identify the impacts of the relationship between environmental performance, the socioeconomic position and SCEM. Then, we construct the SCEM indicator using variables and statistical results from the second step.

Efficiency is measured in terms of either pollutant emission or a combination of output and pollution. This is done by developing the three-step model. The first step estimates the balanced growth indicator (BGI) for each of the eight Asian countries. Balanced growth is evaluated in terms of GDP and CO_2 emission. This step is carried out empirically by an output-oriented directional data envelopment analysis (DEA). The estimated BGI is then applied in the second step. We use the Tobit model to identify the factors affecting BGI. Based on the SCEM framework (Figure 2.1), the three variables representing the environmental management capacities of governments, firms and citizens are selected and used in the Tobit model. These variables and their elasticities are estimated in the second step and used in the third step to construct the SCEM indicator.

Step 1: measuring environmental performance

We begin by measuring environmental performance. There are several ways to quantify such performance. For example, in the case of sulphur dioxide (SO_2), environmental performance is typically measured in terms of SO_2 emissions or SO concentrations. Although these measures are common and frequently used, they may underestimate performance in industrial areas and overestimate it in rural areas. In addition, if two different areas produce the same level of SO_2 emission, then environmental performance will be estimated to be the same, even though other factors (such as monetary value of industrial output, size of the workforce and the value of capital) are different.

One way to overcome such limitation of conventional performance measurement is to estimate efficiency as an environmental performance measurement. The basic idea behind the efficiency approach is presented in Figure 2.2. In this figure, y is the level of desirable output (e.g., GDP, firm's output) and b is the level of undesirable output (e.g., SO_2 and CO_2 emissions). However, b can be interpreted as an unavoidable polluting byproduct resulting from the production of desirable output. Thus, in order to produce a desirable output, it is necessary to produce a certain amount of undesirable output at the same time. A production possibility frontier (PPF) indicates the maximum producible level of desirable output, given the level of undesirable output. Any combination of desirable and undesirable outputs (y, b) is possible under this frontier.

Three different types of efficiencies can be measured in terms of PPF. First, assume that the combination of desirable and undesirable outputs is (y_R, b_R), as shown in Figure 2.2. This combination is inefficient because one can increase the desirable output up to $\hat{y}$ without increasing the undesirable output. This output-oriented efficiency is known as technical efficiency, defined by the distance between the observed level of desirable

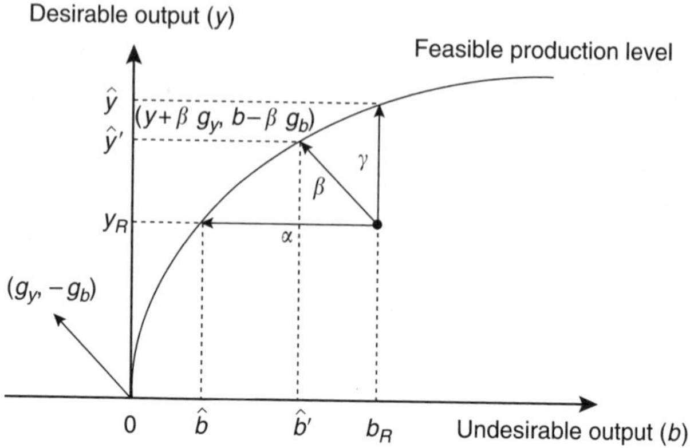

Figure 2.2 Directional distance function

output (y_R) and the maximum producible level ($\hat{y}$). This is denoted as γ in Figure 2.2.

Second, efficiency can be measured in terms of the undesirable output. We continue to assume the same combination of desirable and undesirable outputs (y_R, b_R). In Figure 2.2, one can decrease the level of undesirable output up to $\hat{b}$ without decreasing the output. We define this input-oriented efficiency as EE. This is illustrated by the difference between the observed level of undesirable output ($\hat{b}$) and the minimum feasible level (b_R) in Figure 2.2.

Third, efficiency can also be measured in terms of both desirable and undesirable outputs, represented as β in Figure 2.2. Given the combination of desirable and undesirable outputs (y_R, b_R), one can simultaneously increase the desirable output up to $\hat{y}'$ and decrease the undesirable output up to $\hat{b}'$. This efficiency is denoted as the balanced growth indicator in Watanabe and Tanaka (2006). In order to be consistent with technical efficiency and EE, we denote the third efficiency as BGE.

Since efficiency is measured in terms of the distance between observed and PPF, a small value indicates greater efficiency. The value becomes 0 if the observation lies on the PPF. We modified this efficiency score so that it is truncated from low to high efficiency. Thus, in our modified efficiency score, 0 indicates the least efficiency and 1 indicates the highest efficiency.

Efficiency can be measured empirically using linear programming software such as GAMS and OnFront. Technical resources for theoretical foundations and a linear programming algorithm are presented in the appendix.

Step 2: the relationship between SCEM and environmental performance

Once environmental performance has been estimated, the next step is to identify the impacts of the relationship among environmental performance, the socioeconomic position and SCEM. Specifically, this step evaluates the impacts of the socioeconomic condition and SCEM on environmental performance using the following model:

$$u_{it} = f(G_{it}, F_{it}, C_{it}, X_{it}) \tag{2.1}$$

where u_{it} is the environmental performance for unit i in period t. Environmental performance can be measured by the environmental efficiency or BGE presented on page 30. The unit indicates an economic decision-making unit, depending on the data. If the data is province-specific aggregated data, the unit will be the province. If the data is a firm-level survey, the unit will be the firm. G_{it}, F_{it} and C_{it} represent the capacities of governments, firms and citizens, respectively. Finally, X_{it} is a vector of the socioeconomic situation. In the simplest form, equation (2.1) can be estimated by the following linear model:

$$u_{it} = \beta_0 + \beta_1 G_{it} + \beta_2 F_{it} + \beta_3 C_{it} + \beta_4 X_{it} + \varepsilon_{it} \tag{2.2}$$

where β's are parameters to be estimated and ε_{it} is a disturbance term capturing random noise.

Since environmental performance, estimated in the first step, has a discrete jump at 0 and 1 (i.e., the value always falls between 0 and 1) the parameter estimates of equation (2.2) using the ordinary least square (OLS) method results in baseness and inconsistency (Greene, 2003). Thus, it is necessary to consider the censoring of the dependent variable. One of the most commonly used applications for such censored data is the Tobit model (also referred to as a censored regression). The Tobit model is given as:

$$u_{it} = \begin{cases} \beta_0 + \beta_1 G_{it} + \beta_2 F_{it} + \beta_3 C_{it} + \beta_4 X_{it} + \varepsilon_{it} & \text{if } 0 < \text{RHS} < 1 \\ 0 & \text{if } \text{RHS} \leq 0 \\ 1 & \text{if } \text{RHS} \geq 1 \end{cases} \tag{2.3}$$

where RHS denotes the right-hand side of equation (2.2). This model has been widely used in various fields, including economics, and can be estimated using most of the major econometric software such as Eviews, LIMDEP, and STATA.

Once the Tobit model is estimated and the effects of the socioeconomic situation and SCEM are found to be statistically significant, we compare the contributions of these variables to environmental performance by using elasticity. Elasticity is defined as an incremental percentage change in one variable – in this case, environmental performance – due to a 1 per cent increase in another variable. For example, the elasticity of environmental performance with respect to the capacity of a government can be estimated by $\frac{\partial u}{\partial G} \cdot \frac{G}{u}$, where $\frac{\partial u}{\partial G}$ is the coefficient for the capacity of the government, estimated by the Tobit model.

By estimating the elasticity for each of the socioeconomic condition or SCEM variables, we can reveal the degree of contribution to environmental performance among SCEM and socioeconomic conditions. Further, the estimated elasticities are also used in the next step to construct the SCEM indicator.

Step 3: development of the SCEM indicator

Once the environmental management capacity variables of the three actors are estimated and shown to be significant from the Tobit model, these variables and elasticities are used to construct the SCEM indicator. Following Tanaka and Watanabe (2005), the SCEM indicator is calculated by the following weighted average of the capacities of governments, firms and citizens:

$$SCEM_{it} = \omega_g \tilde{G}_{it} + \omega_f \tilde{F}_{it} + \omega_c \tilde{C}_{it} \qquad (2.4)$$

where $SCEM_{it}$ is the value of the SCEM indicator for country i in year t. $\tilde{G}_{it}$, $\tilde{F}_{it}$ and $\tilde{C}_{it}$ are the normalized capacities of governments, firms and citizens, respectively. The capacity of governments is normalized by $\tilde{G}_{it} = G_{it}/\bar{G}_{it}$, where $\bar{G}_{it}$ is the maximum value of G_{it} in the data set. Moreover, similar normalization is performed for firms and citizens. The 'weights' of the three actors are ω_g, ω_f and ω_c. The weights are determined by the given estimated elasticities from the Tobit model. We normalized the weights such that the sum of elasticities take the value of 1 ($\omega_g + \omega_f + \omega_c = 1$) by using the equation $\omega_f = e_j(e_g + e_f + e_c)$ for $j = G$, F, C. By normalizing all the variables and weights on the right-hand side of equation (2.4), we obtain the convenient and intuitive indicator of the SCEM taking a value between 0 and 1.

Applications of SCEM indicator development

We now present two empirical applications. The first application develops the SCEM indicator in terms of CO_2 emissions in eight countries in east and south-east Asia. This application reveals how the development process and the contributing actors are different among countries. The second application pertains to the domestic scale: SO_2 emissions from industrial sources in China. By focusing only on one country, this application reveals further details in SCEM development.

Application 1: CO_2 emissions in eight asian countries

In this example, the empirical procedure developed in the previous section is applied to CO_2 emissions during 1995–2003 in eight Asian countries – China, Indonesia, Japan, Malaysia, the Philippines, South Korea, Thailand and Malaysia. Environmental performance is measured by BGE, which is determined by efficiency in terms of both desirable and undesirable outputs. Additional details of BGE are presented in the appendix.

Step 1

We first estimate the BGE for the eight Asian countries in the period 1995–2003 using a directional output distance function. In this framework, each country produces two outputs using three inputs. The two outputs constitute GDP as the desirable output and CO_2 as the undesirable output. Labour, capital, and energy consumption are the three inputs.

The data for labour, capital and GDP are gathered from World Development Indicators (2005). Data for energy consumption are obtained from Statistics on Energy Balances by the International Energy Agency/Organization for Economic Cooperation and Development (IEA/OECD) (2005a, 2005b). Data for CO_2 emission are obtained from statistics on total CO_2 emissions from the Carbon Dioxide Information Analysis Center (CDIAC) (2005).

Table 2.1 depicts the estimated BGE in the eight Asian countries between 1995–2003. Overall, the BGE increases by more than 50 per cent in Asia. As expected, Japan is estimated to have the highest BGE score. Although many of the countries exhibit significant increases, the BGE score is low and almost constant for China and Vietnam. This implies that these emerging countries have made little progress toward sustainable development during the estimation period.

Table 2.1 Estimated balanced growth efficiency in eight Asian countries (1995–2003)

| Year | China | Indonesia | Japan | Balanced growth efficiency | | | | | Total |
				Malaysia	Philippines	S. Korea	Thailand	Vietnam	
1995	1.00	0.13	0.95	0.12	0.52	0.25	0.13	0.19	0.41
1996	0.09	0.13	0.96	0.13	0.41	0.26	0.12	0.15	0.28
1997	0.08	0.13	0.99	0.13	0.37	0.29	0.13	0.20	0.29
1998	0.09	0.79	1.00	0.30	0.65	1.00	0.71	0.19	0.59
1999	0.10	1.00	0.98	0.81	0.80	0.92	0.73	0.22	0.69
2000	0.10	0.53	0.98	0.20	0.62	0.88	0.47	0.14	0.49
2001	0.10	0.49	1.00	0.66	0.80	0.89	0.40	0.12	0.56
2002	0.09	0.54	1.00	0.68	0.90	0.99	0.42	0.11	0.59
2003	0.09	0.70	1.00	1.00	1.00	1.00	0.36	0.11	0.66
Mean	0.19	0.49	0.98	0.45	0.67	0.72	0.38	0.16	0.51

Step 2

We now evaluate the effects of the socioeconomic condition and actor capacities on environmental performance by using a censored regression model. Accordingly, EE for province i in year t is explained by the following model:

$$BGE_{it} = \beta_0 + \beta_1 GOV_{it} + \beta_2 FIRM_{it} + \beta_3 CITIZEN_{it}$$
$$+ \beta_4 GDPPER_{it} + \beta_5 INDONESIA + \beta_6 MALAYSIA \qquad (2.5)$$
$$+ \beta_7 THAILAND + \beta_8 D1998 + \beta_9 D1999$$

where the first three variables *GOV*, *FIRM* and *CITIZEN* are the capacities of governments, firms, and citizens, respectively. *GOV* is measured by the number of environmental treaties and multilateral agreements per million people. *FIRM* is represented by ISO 14001 certifications over industry value added. *CITIZEN* is obtained from the HDI. The socioeconomic indicator is considered to be *GDPPER* (i.e., gross domestic product per capita). The last two variables are dummies to be constructed from the country and year perspectives. βs are the parameter coefficients to be estimated. All variables are obtained from international organization publications and databases. Descriptive statistics of the independent variables are summarized in Table 2.2.

The data for environmental treaties are gathered from the Center for International Earth Science Information Network (CIESIN) and the Environmental Treaties and Resource Indicators (ENTRI) database. The data on multilateral environmental agreements are obtained from Earthtrends Environmental Information searchable databases under the management of the World Resource Institute (WRI), based on their collection from each corresponding organization. The data for ISO certification is adopted from annual ISO surveys (2001–04) conducted by the International Organization for Standardization (ISO). The HDI is collected from annual Human Development Reports (1994–2005) prepared by the United Nations Development Programme (UNDP). Since HDI is not reported for the year 1996, we use the data reported in 1995 for the year 1996 and that reported in 1994 for the year 1995, in the estimation. The denominator variables (population and industry value added) and the socioeconomic background variable – gross domestic product per capita – are obtained from the same sources as in the first stage estimation (i.e., World Development Indicators).

Table 2.3 depicts the estimated coefficients and elasticities for the Tobit model. The results show that the estimated model fits the data

Table 2.2 Variables and descriptive statistics for the Tobit model (Application 1)

Variable	Description	Unit	Mean	Std Dev.	Min	Max
GOV	Capacity of governments, measured in terms of total environmental treaties and multilateral agreements per million people	Number of agree-ments over million population	1.20	0.98	0.08	3.79
FIRM	Capacity of firms, measured in terms of the number of ISO 14001 per billion dollars in value added	Number of certifica-tions over billion constant 2000 US$	2.73	3.03	0.00	11.90
CITIZEN	Capacity of citizens, measured in terms of Human Development Indicator (HDI)	Interval between 0 and 1	0.77	0.96	0.56	0.94
GDPPER	Gross domestic product per capita	Billion constant 2000 US$ per million population	6.58	11.69	0.30	38.20
INDONESIA	1 if in Indonesia		0.13	0.33	0.00	1.00
MALAYSIA	1 if in Malaysia		0.13	0.33	0.00	1.00
THAILAND	1 if in Thailand		0.13	0.33	0.00	1.00
D1998	1 if in 1998		0.13	0.33	0.00	1.00
D1999	1 if in 1999		0.13	0.33	0.00	1.00

plausibly well. All independent variables are statistically significant at either 1 per cent, 5 per cent, or 10 per cent levels. In addition, the joint null hypothesis is rejected ($\chi^2 = 72.84$). Results indicate that both socioeconomic conditions (*GDPPER*) and SCEM variables (*GOV, FIRM* and *CITIZEN*) are significant, implying that they affect environmental performance with statistical significance. The results also indicate that the SCEM variables exert greater influence

Table 2.3   Estimated coefficients from the Tobit model (Application 1)

Independent variable	Coefficient	Std Error	Elasticity
Intercept	−0.52	0.33	
GOV	0.09 ***	0.02	0.22
FIRM	0.05 ***	0.01	0.24
CITIZEN	0.83 *	0.45	1.26
GDPPER	0.01 ***	0.01	0.16
INDONESIA	0.28 ***	0.09	
MALAYSIA	−0.10	0.09	
THAILAND	−0.18 *	0.09	
D1998	0.21 **	0.08	
D1999	0.28 ***	0.08	
n	72		
χ^2	72.84		

Notes: One, two, and three asterisks indicate statistical significance at 1 per cent, 5 per cent, and 10 per cent levels respectively.

on environmental performance than the socioeconomic condition, implying the importance of social capacity development of environmental management in the eight Asian countries. Among the three capacities, *CITIZEN* has the largest elasticity and is considerably elastic. *GOV* and *FIRM* exhibit nearly the same levels of elasticities.

Step 3

The third step develops the SCEM indicator based on the results obtained from previous steps and from equation (2.4). Figure 2.3 reports the estimated SCEM indicator for the eight Asian countries for the period between 1995 and 2003 and shows a significant increase in the SCEM for all of the eight countries. Japan is estimated to be the highest in the SCEM ratings among these countries. The SCEM in Japan increases from 0.77 to 0.88, accounting for an increase of 15 per cent. The second highest SCEM is observed in Korea. Overall, the ranking of the SCEM indicator is mostly consistent with the degree of economic development among these countries. Although China has one of the lowest SCEM values, it exhibits the most rapid increase, which accounts for nearly 40 per cent during the estimation period.

The estimated capacities of governments, firms and citizens are shown in Figures 2.4, 2.5 and 2.6, respectively. These figures reveal the capacity of firms (the number of ISO certifications per billion dollars) as a

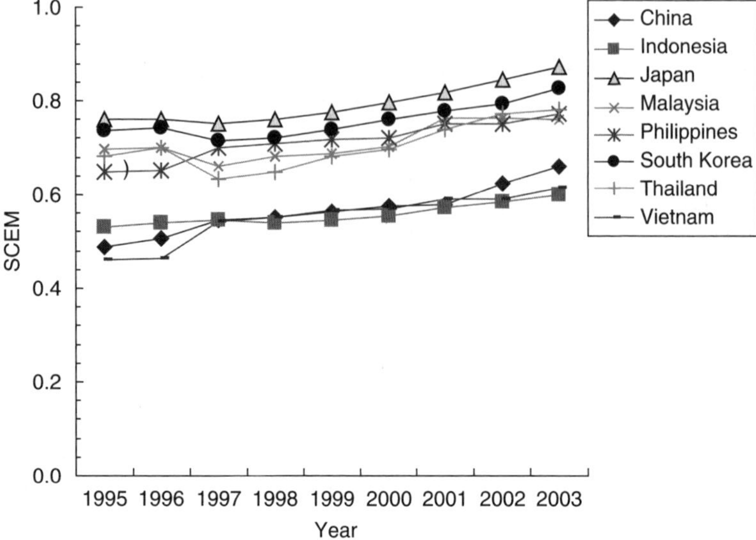

Figure 2.3   Estimated SCEM indicator in eight Asian countries (1995–2003)

primary source of SCEM development. As shown in Figure 2.5, firms in most countries enhanced their capacities rather rapidly during the estimation period. Particularly high increases are observed in Thailand and Japan. Although the lowest score is observed in China until 2001, the country increases its capacity rather rapidly after 2002 and keeps pace with other countries. The capacity of citizens (HDI) is almost constant, particularly after 1997. Although its elasticity is relatively large, capacity enhancement is slow in all the eight Asian countries considered in this study. Finally, the capacity of government is observed to be decreasing in all countries except Japan and Malaysia, albeit rather slowly. This reflects the recent stagnation of international agreements pertaining to CO_2 emission reduction and other global environmental issues.

Application 2: SO_2 emissions from industrial sectors in China

In this example, the empirical procedures presented in the previous section are applied to SO_2 emissions from industrial sectors (electricity, mining, and manufacturing sectors) in China during the period between 1994 and 2002. Environmental performance is measured by EE in terms of SO_2 emission; i.e., the difference between the minimum feasible level

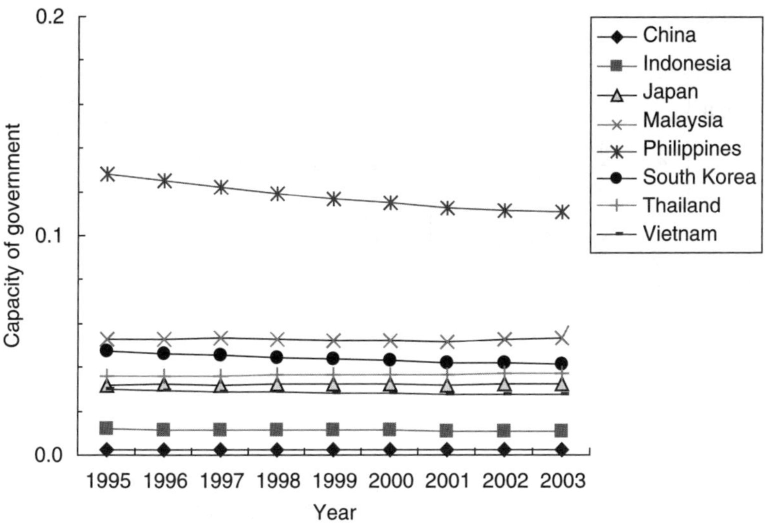

Figure 2.4 Estimated capacity of government in eight Asian countries (1995–2003)

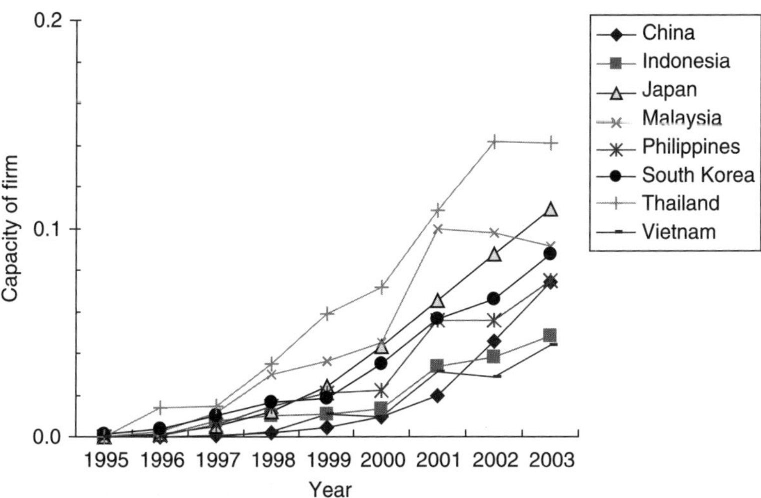

Figure 2.5 Estimated capacity of firms in eight Asian countries (1995–2003)

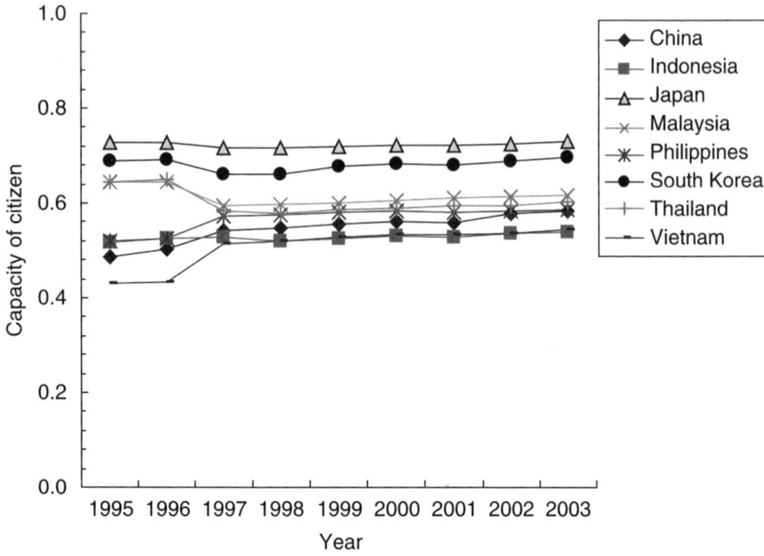

Figure 2.6   Estimated capacity of citizen in eight Asian countries (1995–2003)

of emission and the observed level of emission, at the given level of GDP. This corresponds to the distance α in Figure 2.2.

Step 1

We begin by estimating EE, described above, for 30 provinces in China[1] For this purpose, we set up a province-level aggregated production function consisting of two outputs and three inputs. The outputs include both desirable and undesirable outputs. The desirable output is measured in terms of the gross industrial output value. The undesirable output – pollutants as byproducts of industrial operation – is measured in terms of the level of SO_2 emission from the industrial sectors. With regard to the inputs of the model, we select the size of the labour force, capital and coal as inputs. Capital is measured in terms of the net value of fixed assets. Labour is measured in terms of the number of staff and workers. Coal is measured in terms of coal consumption as an input material. These variables are obtained from the *China Statistical Yearbook*.

Table 2.4 depicts the estimated EE in China between 1994–2002. The national average of the efficiency score is estimated to be 0.77, with some fluctuation during the estimation period. Specifically, efficiency

Table 2.4   Estimated environmental efficiency in China (1994–2002)

Year	Coast	Environmental efficiency Central	West	Nation
1994	0.79	0.80	0.80	0.80
1995	0.71	0.79	0.80	0.76
1996	0.72	0.80	0.80	0.77
1997	0.74	0.81	0.80	0.77
1998	0.69	0.74	0.77	0.73
1999	0.75	0.77	0.77	0.76
2000	0.74	0.78	0.74	0.75
2001	0.79	0.79	0.78	0.79
2002	0.83	0.79	0.78	0.81
Mean	0.75	0.79	0.78	0.77

gradually decreases until the year 2000. However, the highest efficiency score is achieved in 2002, the last year of our estimation period.

Step 2

In the second step, using the EE estimated in the first step, we evaluate the effects of the socioeconomic condition and SCEM on EE using the censored regression model. Specifically, EE for province i in year t is explained by the following model:

$$BGE_{it} = \beta_0 + \beta_1 GPPC_{it} + \beta_2 IRATIO_{it} + \beta_3 GOV_{it} + \beta_4 FIRM_{it}$$

$$+ \beta_5 CITIZEN_{it} + \beta_6 CENTRAL + \beta_7 WEST + \beta_8 TIME \qquad (2.6)$$

where the first two independent variables represent socioeconomic conditions. *GPPC* denotes gross provincial product per labour and *IRATIO* represents the ratio of industrial sector. The next three variables *GOV*, *FIRM* and *CITIZEN* are the capacities of governments, firms and citizens, respectively. *GOV* is measured in terms of the number of monitoring stations. *FIRM* is represented by the SO_2 removal rate (ratio of total SO_2 removed to total SO_2 emission). *CITIZEN* is denoted by the number of environmental disputes. *CENTRAL* and *WEST* are dummy variables for provinces in the central and western regions. *TIME* is a time trend capturing time-varying effects. βs are the parameter coefficients to be estimated. All variables are obtained from the *China Statistical Yearbook* and the *China Environmental Yearbook* between the years 1995 and 2003.

Table 2.5   Variables and descriptive statistics for the Tobit model (Application 2)

Variable	Description	Unit	Mean	Std Dev.	Min	Max
GPPC	Gross provincial product per labour, deflated by GDP index	100 million yuan	0.06	0.03	0.02	0.17
IRATIO	Ratio of industrial sector in terms of output value	Percentage terms	0.36	0.15	0.08	1.01
GOV	Capacity of government, measured by the number of monitoring stations	Counts	1.26	0.60	0.00	4.00
FIRM	Capacity of firm, measured by the rate of SO_2 removal (ratio of SO_2 removal and SO_2 emission)	Percentage terms	0.21	0.16	0.01	0.68
CITIZEN	Capacity of citizen, measured by the number of citizen groups complaining over environmental issues	Counts	734.76	798.20	1.00	5530.00
CENTRAL	= 1 if in Central region		0.31	0.46	0.00	1.00
WEST	= 1 if in Western region		0.24	0.43	0.00	1.00
TIME	Time trend (1994 = 1)	Year	5.00	2.59	1.00	9.00

Descriptive statistics of the independent variables are summarized in Table 2.5.

Table 2.6 presents the estimated results. Overall, the model fits the data rather well. All independent variables except *CITIZEN* are statistically significant at the 1 per cent level. In addition, a chi-square statistic (123.08) strongly rejects the null hypothesis that all coefficients are simultaneously 0. As expected, the signs of the socioeconomic condition (*GPPC* and *IRATIO*) and environmental management

Table 2.6   Estimated coefficients from the Tobit model (Application 2)

Independent variable	Coefficient	Std Error	Elasticity
Intercept	0.21 ***	0.06	
GPPC	4.69 ***	0.53	0.36
IRATIO	0.68 ***	0.11	0.32
GOV	0.08 ***	0.02	0.12
FIRM	0.25 ***	0.07	0.07
CITIZEN	< 0.01	0.00	0.08
CENTRAL	0.16 ***	0.03	
WEST	0.10 ***	0.03	
TIME	−0.03 ***	0.01	
n	190		
χ^2	123.08		

Notes: Three asterisks indicate statistical significance at 1 per cent.

capacities (*GOV, FIRM* and *CITIZEN*) are shown to be positive. This indicates that both the socioeconomic conditions and SCEM are positive inducements for efficiency in SO_2 emissions from the industrial sector in China. The capacity of citizens (*CITIZEN*) is not significant at any statistical level. This may imply that the citizens' capacity is not sufficiently well developed to induce an increase in environmental efficiency and a consequent abatement of the SO_2 pollution problem in China.

Moreover, Table 2.6 presents the estimated elasticities of the socioeconomic position and SCEM. Two socioeconomic conditions are relatively elastic: the elasticities of *GPPC* and *IRATIO* are estimated to be 0.36 and 0.32, respectively. These indicate that an increase of 1 per cent in *GPPC* and *IRATIO* result in an increase in EE by 0.36 per cent and 0.32 per cent, respectively. The elasticities of SCEM variables are relatively small; an increase of 1 per cent in the capacities of governments, firms and citizens will enhance the EE by 0.12 per cent, 0.07 per cent and 0.08 per cent, respectively. Among the three actors of the SCEM, the capacity of governments is the most elastic. This indicates that capacity development in governments is the most effective in enhancing efficiency in terms of SO_2 emission from the industrial sectors. The capacity of firms is estimated to have the smallest elasticity among the capacities considered in this study. However, its elasticity is significant whereas that of citizens is not.

Step 3

The third step develops the SCEM indicator based on the results obtained from the first and second steps. Specifically, the SCEM indicator is calculated as follows:

$$SCEM_{it} = \omega_g \tilde{G}_{it} + \omega_f \tilde{F}_{it} \qquad (2.7)$$

where $\tilde{G}_{it}$ and $\tilde{F}_{it}$ are the capacities of governments and firms, respectively, which are normalized such that they take a value between 0 and 1. The weights of capacities are ω_g and ω_f, derived from the elasticities reported in Table 2.6. These weights are adjusted such that the sum of the weights is 1. This indicator is a modified version of equation (2.4). The only difference is that the capacity of citizens is not included here due to the insignificance estimated in the Tobit model.

Figure 2.7 presents the estimated SCEM indicator and the observed levels of SO_2 emission between 1994–2000. These values are normalized so that they take a value between 0 and 1 for comparison purpose. As Figure 2.7 shows, the SCEM indicator increases from 0.27 in 1994 to 0.37 in 2002, accounting for a growth of 37 per cent. Thus, environmental management capacity development has become fairly well developed during the estimation period. In the same period, SO_2 emission decreased by about 10 per cent. It is a modest increase under rapid

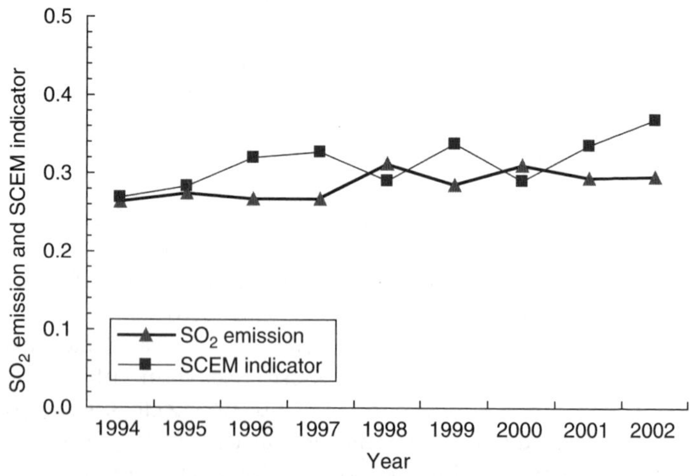

Figure 2.7 SO_2 emissions and SCEM indicator in China (1994–2002)

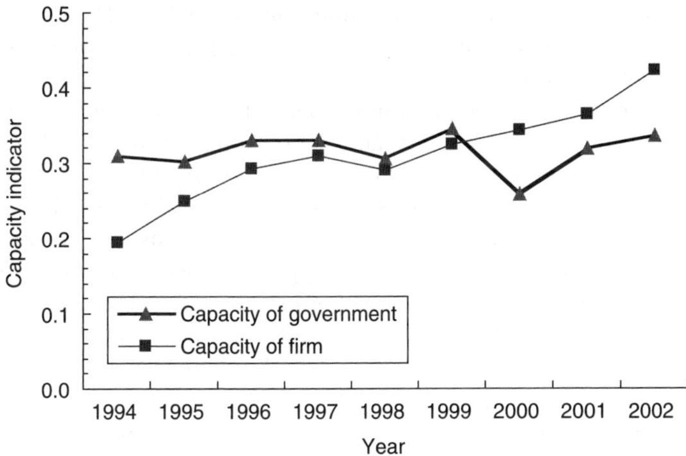

Figure 2.8   Capacities of government and firms in China (1994–2002)

economic development during the period, and is significantly due to the SCEM development during the period.

Finally, Figure 2.8 illustrates the changes in the capacities of governments and firms during the same period. The figure illustrates that the SCEM is somewhat increased during the estimation period, and also that there is significant development in the capacity of government. It increases from 0.19 in 1994 to 0.42 in 2002, accounting for an increase of 118 per cent. In contrast, the capacity of governments is estimated to be nearly constant during the estimation period. Thus, although the SCEM is developed by nearly 40 per cent during 1994–2002, such capacity development is mostly due to a rapid enhancement of the capacity of firms. Further development in the SCEM may require enhancing the capacity of government as well as that of firms.

Summary and conclusion

This chapter develops an indicator of SCEM using the three-step modelling approach. The first step estimates environmental performance by measuring efficiency. We present two different efficiency measures. The first measure is EE, which focuses only on the emission of pollutants. The second is BGE, which estimates the efficiency of sustainable economic development by dealing with both output and pollutant emission. The estimated environmental performance is then used in the second step to identify the impacts of the relationship between environmental performance, the socioeconomic conditions and SCEM. Then, we construct the SCEM indicator using the variables and statistical results obtained from the second step.

Based on the empirical procedures discussed on pp. 30–4, we report two empirical applications. The first application compares the SCEM and its development among eight Asian countries between 1995–2003. We evaluate the environmental performance of each country by the BGE and the efficiency measurement in terms of GDP and CO_2 emission. We observe a significant increase in environmental performance in the eight Asian countries. The Tobit model reveals both the SCEM and socioeconomic conditions to be a significant source of such performance enhancements. We then construct the SCEM indicator for the eight Asian countries. The indicator reveals that all eight countries develop SCEM fairly well. Such a development in SCEM is mostly due to the capacity enhancement of firms. In contrast, the capacities of governments and citizens remain almost constant during the estimation period.

We also report another empirical example, which focuses on industrial SO_2 emission in China between 1994–2002. In this example, environmental performance is defined by EE in terms of SO_2 emissions from the industrial sectors in each of the 30 provinces. The directional distance output function reveals that although the EE decreases marginally during the estimation period, the highest score is observed in the last year of the period of estimation (2002). As observed in the first example, the estimated environmental performance is influenced by both the socioeconomic condition and SCEM. However, the capacity of citizens is estimated not to be significant at any statistical levels. This may imply that the capacity of citizens is not very developed and therefore has no significant influence on the abatement of environmental pollution in China. Finally, we construct the SCEM using the capacities of governments and firms. The capacity of citizens is not included because it is statistically insignificant. Our results reveal that the SCEM is fairly well increased between 1994–2002. Further, they indicate that this is mostly due to a rapid enhancement of the capacity of firms. Although this capacity increases by more than 100 per cent, the capacity of governments is estimated to be almost constant during the estimation period.

Before concluding this chapter, several limitations need to be identified. First, because the relationship between environmental performance and the SCEM does not need to be linear, a more flexible functional form may be appropriate in the Tobit model. For example, a translog or quadratic form may produce more accurate measurements of such a relationship. Further, interactions among governments, firms and citizens may need to be included as interaction terms in the model. Second, this chapter employed only three variables – governments, firms and citizens. Although this simplicity significantly facilitates indicator development, it may be relatively easy to investigate each of the three actors for further details. As Figure 1.6 in Chapter 1 illustrates, these actors typically consist of the following three factors: (1) policy and measures; (2) organizational resources; and (3) knowledge and technology. However, the more capacity included in the model, the more likely is the model to have multicollinearity (i.e., correlation among independent variables). This may result in biased and inefficient estimates.

Thus, the SCEM indicator can be used as a convenient overview of the SCEM development with readily available data. Further investigation of each actor can be implemented using another approach with factor-specific data for the three actors. The next chapter introduces one such methodology, the actor–factor analysis.

Appendix: directional output distance function

This appendix presents the technical details involved in the estimation of the directional distance function, derived from Färe and Grosskopf (2004), Watanabe (2004), and Watanabe and Tanaka (2006).

We first denote a vector of inputs by $x = (x_1, \ldots, x_N) \in \Re_+^N$. There are two types of outputs – desirable output (e.g., GDP) and undesirable output (e.g., SO_2 emission) – which are denoted as $y = (y_1, \ldots, y_M) \in \Re_+^M$ and $b = (b_1, \ldots, b_I) \in \Re_+^I$, respectively. The relationship between input and output is represented by an output set:

$$P(x) = \{(y, b) : x \text{ can produce } (y, b)\}, \quad x \in \Re_+^N \tag{2.A.1}$$

The output set is assumed to have the following properties. The first is 'null-jointness', which implies that the desirable output cannot be produced without simultaneously producing the undesirable output:

$$(y, b) \in P(x) \quad \text{and} \quad b = 0, \quad \text{then } y = 0 \tag{2.A.2}$$

The second and third properties relate to the production technology of desirable and undesirable outputs. The second assumption is referred to as the weak disposability of the undesirable output:

$$(y, b) \in P(x) \quad \text{and} \quad 0 \leq \theta \leq 1, \quad \text{then} \quad \theta(y, b) \in P(x) \tag{2.A.3}$$

This indicates that it is impossible to reduce the undesirable output without reducing the desirable output. The third assumption is known as the strong disposability of the desirable output:

$$(y, b) \in P(x) \quad \text{and} \quad y^0 \leq y, \quad \text{then} \quad (y^0, b) \in P(x) \qquad (2.A.4)$$

This assumption implies that it is possible to reduce the desirable output without reducing the undesirable output. Thus, within this model, there exists an asymmetry in the properties of desirable and undesirable outputs. Figure 2.2 depicts the output set, $P(x)$, for a case comprising one desirable output and one undesirable output. The output set satisfies the 'null-jointness' property as the function passes through the origin. It is possible to reduce the desirable output (i.e., a vertical downward shift in production is possible) although it is not possible to reduce the undesirable output (i.e., a horizontal shift left is not possible for observations on the PPF).

Given the foundations in equations (2.A.1)–(2.A.4), the directional output distance function is now defined as follows:

$$\vec{D}_o(x, y, b; g_y, g_b) = \max\{\beta : (y + \beta g_y, b - \beta g_b) \in P(x)\} \qquad (2.A.5)$$

The value of β represents the distance between the observation (y, b) and a point on the PPF, $(y + \beta g_y, b - \beta g_b)$. A direction vector, $g = (g_y, -g_b)$, determines the direction in which efficiency is measured. The direction vectors for desirable and undesirable outputs are represented by g_y and g_b, respectively. Given the production technology $(p(x))$ and direction vector (g), the directional distance function yields the maximum feasible expansion of the desirable output and the contraction of the undesirable output. If the observation is on the production frontier, then the value of the directional output distance function is zero and the observation is the most efficient. As the value increases, the observation becomes less efficient. Thus, contrary to the distance function, the efficiency of an observation increases as the value of the directional distance function approaches 0. It is possible to measure the efficiency in different directions by changing the value of the direction vector.

The output directional distance function, $\beta^{k'}$, for the kth observation is obtained by solving the following maximization problem:

$$\vec{D}_o(x^{k'}, y^{k'}, b^{k'}; g_y, g_b) = \max \beta^{k'} \qquad (2.A.6)$$

$$\text{s.t.} \ \sum_{k=1}^{K} z_k y_{km} \geq y_{k'm} + \beta^{k'} g_{ym} \quad m = 1, \ldots, M \qquad (2.A.7)$$

$$\sum_{k=1}^{K} z_k y_{ki} = b_{k'i} - \beta^{k'} g_{b_i} \quad i = 1, \ldots, I \tag{2.A.8}$$

$$\sum_{k=1}^{K} z_k x_{kn} \leq x_{k'n} \quad n = 1, \ldots, N \tag{2.A.9}$$

$$\sum_{k=1}^{K} z_k = 1 \tag{2.A.10}$$

$$z_k \geq 0 \quad k = 1, \ldots, K, \tag{2.A.11}$$

where z_k is the weight of the kth observation. Note that equations (2.A.7) and (2.A.8) represent the strong disposability of desirable output and the weak disposability of undesirable output, respectively. Equation (2.A.10) is omitted if CRS is assumed.

Note

1   Tibet is excluded from this analysis due to data limitation.

References

Armitage, J. and G. Schramm (1989) 'Managing the Supply and Demand for Fuelwood in Africa', in Schramm G. and J.J. Warford (eds), *Environmental Management and Economic Development* (Washington, DC: World Bank).

CDIAC (Carbon Dioxide Information Analysis Center) (2005) *Global, Regional, and National CO2 Emission Estimates from Fossil Fuel Burning, Cement Production, and Gas Flaring: 1751–2000* (Oak Ridge: CDIAC) http://cdiac.esd.ornl.gov/ftp/ndp030/ (accessed on 1 August, 2006).

CIESIN (Center for International Earth Science Information Network) (2005) *Environmental Treaties and Resource Indicators (ENTRI)* (New York: CIESIN, Columbia University) http://sedac.ciesin.columbia.edu/entri/partySearch.jsp (accessed on 1 August, 2006).

Editorial Board for China Environment Yearbook (1995–2003) *China Environment Yearbook* (Beijing: China Environmental Science Press).

Esty, D.C., L. Marc, S. Tanja and de S. Alexander (2005) *2005 Environmental Sustainability Index: Benchmarking National Environmental Stewardship* (New Haven: Yale Center for Environmental Law and Policy).

Färe, R. and S. Grosskopf (2004) *New Directions: Efficiency and Productivity* (Boston: Kluwer Academic Publishers).

Greene, W. (2003) *Econometric Analysis, 5th edn.* (Englewood Cliffs, NJ: Prentice-Hall).

IEA/OECD (International Energy Agency/Organisation for Economic Co-operation and Development) (2005a) *CO$_2$ Emissions from Fuel Combustion 1971–2003* (Paris: OECD).

IEA/OECD (2005b) *Energy Balances of Non-OECD Countries 2002–2003* (Paris: OECD).

Marlies, G., M. Kaldor and H. Aheier (eds) (2005) *Global Civil Society 2005/6* (London: Sage).

Matsuoka, S. and A. Kuchiki (eds) (2003) *IDE Spot Survey, Social Capacity Development for Environmental Management in Asia – Japan's Environmental Cooperation after Johannesburg Summit 2002* (Chiba: IDE-JETRO).

National Bureau of Statistics of China (1995–2003) *China Statistical Yearbook* (Beijing: China Statistical Press).

State Statistical Bureau (1995) *China Statistical Yearbook 1995* (Beijing: China Statistical Publishing House).

State Statistical Bureau (1996) *China Statistical Yearbook 1996* (Beijing: China Statistical Publishing House).

State Statistical Bureau (1997) *China Statistical Yearbook 1997* (Beijing: China Statistical Publishing House).

State Statistical Bureau (1998) *China Statistical Yearbook 1998* (Beijing: China Statistical Publishing House).

State Statistical Bureau (1999) *China Statistical Yearbook 1999* (Beijing: China Statistical Publishing House).

State Statistical Bureau (2000) *China Statistical Yearbook 2000* (Beijing: China Statistical Publishing House).

State Statistical Bureau (2001) *China Statistical Yearbook 2001* (Beijing: China Statistical Publishing House).

Tanaka, K. and M. Watanabe (2005) 'Social Capacity for Environmental Management and its Indicator Development: EE Approach', Paper presented at the Japan Society for International Development, Kobe, Japan, 26–27 November (in Japanese).

UNEP/WHO (1996) *Air Quality Management and Assessment Capabilities in 20 Major Cities* (Nairobi: United Nations Environment Programme).

UNDP (United Nations Development Programme) (1994–2005) *Human Development Report 1994* (New York: UNDP).

Watanabe, M. (2004) 'A New Method of Evaluating China's Industrial Efficiency', Paper presented at the Japan Society for International Development, Nagoya, Japan, 29–30 November (in Japanese).

Watanabe, M. and K. Tanaka (2006) 'Efficiency Analysis of the Chinese Industry: A Directional Distance Function Approach', unpublished.

World Bank (2005) *World Development Indicators 2005* (Washington, DC: World Bank).

World Resources Institute (2005) *World Resources 2005: The Wealth of the Poor: Managing Ecosystem to Fight Poverty* (Washington, DC: World Resources Institute).

3
Actor–Factor Analysis for Social Capacity Development

Kazuma Murakami and Shunji Matsuoka

Capacity assessment standards and framework

Since the beginning of the 1990s, the concept of capacity development in the field of development assistance has come to be used frequently when considering improving the quality of assistance, and the UNDP, CIDA, JICA and other assistance organizations have been advancing the debate on putting it into practice (UNDP, 1998; Lavergne and Saxby, 2001; JICA, 2004). In the face of increasingly serious global environmental problems and considering the need to increase ODA to confront environmental destruction, which is one cause of poverty, attention has inevitably turned to the capacity to deal with environmental issues.

The report of the 1987 World Commission on Environment and Development advocated sustainable development, and Agenda 21, adopted at the 1992 UN Conference on Environment and Development, expressed the need for the capacity to cope with environmental problems. In response to this, discussions have also occurred at the OECD/DAC about the concepts and definitions of capacity in environment and capacity development in environment (CDE) (OECD/DAC, 1999).

In consideration of this, Matsuoka *et al.* (2004), have stated that developing the social capacity for environmental management (SCEM), which is the total capacity of society as a whole, links to the efficient and effective implementation of environmental policies and international environmental cooperation. This overall capacity is comprised of the capacities of governments, firms and citizens to cope with environmental problems and the mutual relationships created between these actors.

SCEM is the latent capacity to maximize 'environmental governance' (Matsushita, 2002) that advances environmental management through the appropriate participation of governments, firms, citizens and society as a whole. Given this, it is necessary to create as much capacity as possible for governments, firms and civil society to fill their roles as the bearers of responsibility for the environment as a common public good and provider of services.[1] SCEM is a concept that considers the role sharing of the actors involved in solving environmental problems, and it can be used to develop methods for assessing the capacity required to fulfil those roles.

However, even though the OECD/DAC (1999) and other organizations share a focus on capacity and a recognition of the importance of capacity development, no unified consensus has been reached regarding how to define capacity; abstract, conceptual definitions are all that have been realized. Regarding this, Matsuoka and Honda (2002) state, 'since the concept is broad and abstract, assistance organizations have taken it as an idea, but not enough consideration of the definition or concrete details has occurred'. In addition, even when capacity formation elements have been expressed concretely, as in Janicke and Weidner (1997), Boesen and Lafontaine (1998), and UNDP/GEF (2003) and other organizations, they have not been expressed logically and systematically.

Assessments of capacity are necessary both before and after capacity development efforts. However, effective assessment cannot be achieved if the capacity in question is not made clear and suitable assessment standards established. In order to establish assessment methods for the capacity to handle environmental problems, first, systematic delineation from a logical and empirical perspective of the capacity that is to be assessed is necessary. In this case, the capacity that should be assessed is the capacity to contribute to policy results, and the development of a capacity assessment framework that sets this as the assessment standard is necessary. This investigation process should lead to the proposal of efficient, effective capacity assessment methods.

In this chapter, we focus on the subject of urban air pollution countermeasures. First, we clarify the capacity elements that should be assessed, or, in other words, the capacity assessment standards (factor analysis). Then, we propose a capacity assessment framework using those assessment standards (actor analysis). Finally, we explain the advantages of those capacity assessment methods (the actor–factor matrix) in comparison with other assessment methods.

Setting capacity assessment standards

Background and purpose

UNEP/WHO (1996) has tried to clarify internal structures related to urban air quality management capacity (Figure 3.1). In this case, measurements,

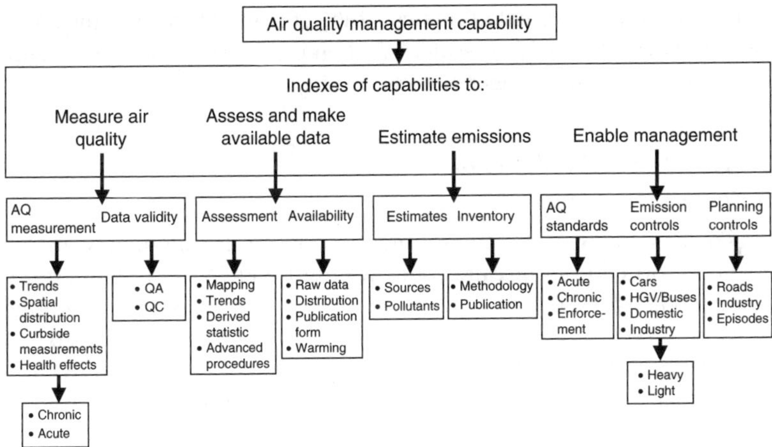

Figure 3.1   Internal structures related to urban air quality management capacity
Source: UNDP/WHO (1996).

data assessments and verifications, emissions source investigations and environmental management are set as four elements of capacity related to air quality management from the point of view of experience, and the air quality management capacities of 20 cities have been assessed and given scores. However, the relative importance and relevance of each element of capacity have not been verified, and the degrees of their contributions to air quality improvement have not been shown. In addition, each element of capacity is given the same maximum potential score of 25 points regardless of its number of component assessment items or its relative importance to air quality improvement. Considering this, doubts exist about the validity of these elements of capacity. In addition to this arbitrary assessment evaluation, the clarity of the concepts in relation to capacity, the measurement methods and other issues suggest that more research is necessary.

In this section, by empirically expressing elements of capacity that contribute to improvement in environmental quality, we seek to develop a practical capacity assessment framework that converts these into assessment measures. Specifically, by identifying elements of capacity related to environmental countermeasures, and empirically showing their relationships to policy results that lead to environmental quality improvements, we will be able to suggest capacity elements that can contribute to environmental quality improvement.

In this section, we detail analysis subjects, analytical methods and data, and describe analysis results. We then propose capacity assessment standards based on the clear analysis results.

Subjects and methods of analysis

Setting analysis subjects

Our specific focus in the environmental countermeasure field is urban air pollution countermeasures, and we use UNEP/WHO (1996) efforts as a reference for comparison. Among air pollution substances, sulphur dioxide (SO_2) is a representative industrial air pollution from factories, workplaces and other fixed emissions sources.

Our analysis period was from 1970 to 2000. According to Harashima and Morita (1995), this was the nascent period for environmental policies (1965~), and, according to Matsuoka and Kuchiki (2003), this was the period when the social capacity for environmental management began to function significantly (1970~). Environmental policies started to be implemented seriously, and their results began to become apparent during these decades. Around the start of this period, according to data from long-term continuous monitoring stations around Japan, annual average SO_2 concentration peaked in 1967, and NO_2 concentration peaked in 1971 before starting to decline.

The subjects of our analysis were the distinct capacities of governments, firms and citizens. This is in keeping with the definition of SCEM offered by Matsuoka *et al.* (2004). According to Ueta (1996), SO_2, as an industrial pollution, is the result of external diseconomy – that is, the failure of the market – and environmental policies based on government intervention, particularly command and control regulations, have had principal roles as countermeasures. Specifically, governments have a certain authority over firms that are polluters, and have advanced air quality improvements by promoting fuel conversion, operation management, new technology development and introduction, and other pollution reducing behaviour. In addition, the so-called 'Pollution Diet' (Kogai-Kokkai), the Japanese national parliamentarians in office around 1970, passed and revised 14 laws related to pollution (1970), established the Japanese Ministry of the Environment (1971) and strengthened other government environmental functions (Harashima and Morita, 1995). For these reasons, we will focus on the capacity of governments as our analysis subject.

Our subject cities are Kitakyushu and Osaka, two ordinance designated cities with different urban structures. As Figure 3.2 shows, the urban

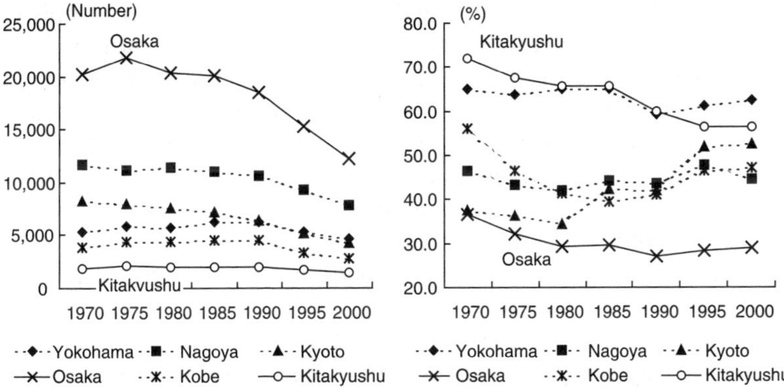

Figure 3.2 Numbers of manufacturing establishments (left), share of value of manufactured goods shipments (manufacturing establishments with 300 or more employees) (right)

Source: Research and Statistics Department, Economic and Industrial Policy Bureau, Ministry of Economy, Trade and Industry (1970–2000).

structures of these two cities are not very similar, with differences in the number of manufacturing workplaces and the proportion of large-scale workplaces (as shown by the ratio of the value of manufactured products shipped from workplaces with over 300 employees). In a comparison of all the ordinance designated cities in 1970, Kitakyushu had the fewest manufacturing workplaces, but its proportion of large-scale workplaces was highest. In contrast, Osaka had the most manufacturing workplaces, but its proportion of large-scale workplaces was lowest. In addition, Osaka had the largest government budget and the greatest number of public employees, while Kitakyushu and Kyoto had the smallest budgets and staffs. Furthermore, Osaka had the highest population and population density and Kitakyushu had the lowest.

These differences in urban structures, especially in industrial structures, resulted in different pollution emission situations and countermeasures for the emissions from factories, workplaces and other fixed emissions sources, particularly for the main pollutant SO_2. The main subjects of regulations in Kitakyushu have been a small number of large firms with extensive technological and financial resources to apply to environmental measures, so transaction costs for communication with these firms have been small. The implementation and efficacy of policies have been achieved through information exchanges, preliminary meetings and other regular meetings (Kitakyushu City, 1998; Katsuhara, 2001; Fujikura, 2002). In contrast, the numerous medium and small firms in

Osaka have had limited technological and financial resources to apply to environmental measures, and transaction costs for communication have been large, so there has been no choice but to seek the implementation and efficacy of policies through the accumulation of government guidance efforts for individual firms (Osaka City, 1994; Fujikura, 2002). Given this, the extent of the transaction costs needed for promoting countermeasures by firms has been a primary factor in the formation of the SO_2 policies of both cities. This resulted in the selection of different policy measures, with Kitakyushu adopting pollution control agreements and Osaka choosing administrative guidance.

For these reasons, we selected Kitakyushu and Osaka as representative examples of applying rational policy measures based on urban structure. We clarified the elements of capacity related to air quality management and contributions to air quality improvement in these two cities with different urban structures and policy measures. Then, through this comparative investigation, we developed conclusions that are common to both cities and that are specific to each.

Analytical methods

We conducted a two-stage empirical analysis in which we first identified elements of capacity related to air quality management and then confirmed that these elements of capacity actually contribute to air quality improvement.

Analysis 1: identification of capacity elements   We conducted factor analysis using data related to government air pollution policies to identify elements of capacity. Factor analysis is a method of identifying multiple common, latent variables that are behind actual phenomena. In this case, several air pollution policies were classified by common factors. The latent commonalities of those classified air pollution policies were derived as the capacity of those policies to be effective. Thus, those factors were shown to be elements of capacity for government air pollution policies. In short, factor analysis makes the latent, abstract concept of capacity visible. This analytical method suited the goals of our research because it also allows the quantitative determination of each factor's contribution to total capacity.

Analysis 2: verification of capacity elements for air quality improvement We verified whether the individual elements of capacity determined in analysis 1, as well as the total capacity of air quality management formed by the combination of weighted capacity elements, contributed to air quality improvement. In other words, we verified the contribution of the derived elements of capacity, individually and as a whole, to policy effectiveness. This also had the result of verifying the existence of the individual elements of capacity and the validity of their weightings.

To verify each element of capacity for air quality improvement, we set the factor scores for the elements derived from the preceding factor analysis as the explanatory variables, and conducted regression analysis with SO_2 concentration as the dependent variable. This clarified the relationship between each element of capacity and air quality.

To verify the total combined capacity of the elements, first, we used the contribution rates of the factor loads as weights to make a weighted average of the factor scores of all the elements obtained through the factor analysis. Thus, we derived the total capacity for air quality management. Then, we conducted regression analysis using this as the explanatory variable with SO_2 concentration as the dependent variable to clarify the total capacity for air quality management and its relationship with air quality. The contribution rates of the factor loads show the degree to which each factor explains the total capacity for air quality management, so they can be taken as the weights of the elements of capacity. Thus, by confirming the fitness of these relationships through regression analysis of the capacity derived from weighted elements of capacity and air quality, which is the result of the policies, the validity of the capacity for air quality management, and, thus, the weightings of the capacity elements are verified.

Data

For our analysis of government subjects, in addition to city governments, we also examined the municipal environmental science laboratories involved in pollution countermeasures that provide scientific data and information for policy development and execution. Specifically, we choose to investigate the Kitakyushu City Institute of Environmental Sciences and the Osaka City Institute of Public Health and Environmental Sciences.

We were able to obtain continuous quantitative data for both cities for the period from 1970 to 2000 regarding each organization's 'input' (personnel, equipment and facilities, funds, information) and 'output' directly related to air pollution policies (Table 3.1). Our subject was data that shows the degree of government activity related to air pollution policies for comparison with SO_2 concentration improvement, which is the 'outcome' of those policies.

Table 3.1 Data resources used at factor analysis

Data	Source

Municipal environmental science laboratories
  Budgets
  Personnel
  Number of research presentations at academic conferences
  Number of articles in academic journals
  Number of specimens analyzed for air pollution
	Report of Kitakyushu City Institute of Environmental Sciences
	Annual Report of Osaka City Institute of Public Health and Environmental Sciences
	The Survey on Wages of Local Government Employees/Ministry of Internal Affairs and Communications

The cities
  Sanitation expenses
  Environmental division personnel
  Number of ambient air pollution monitoring stations
  Number of onsite inspections of factories and workplaces related to air pollution
  Number of cases of air pollution preventing financing funds provided to firms
  Amount of air pollution prevention financing funds provided to firms
	Environment in Kitakyushu City
	White paper on Environment Pollution in Osaka City
	Survey on Wages of Local Government Employees/Ministry of Internal affairs and Communications
	The Annual Report of Local Finances/Minisitry of Home Affairs
	Air Pollution in Japan/Ministry of Environment

For these municipal institutes involved in pollution countermeasures, we were able to gather data on their budgets and personnel (number of people employed×average number of years employed), the number of research presentations at academic conferences, the number of articles in academic journals, and the number of specimens analyzed for air pollution by the institutes in response to requests from government agencies and residents.

For the cities themselves, we gathered data on sanitation expenses, the number of ambient air pollution monitoring stations, environmental division personnel (number of people employed × average number of years employed), the number of onsite inspections of factories and workplaces related to air pollution due to air pollution prevention laws and pollution prevention regulations, citizen complaints and related incidences, the number of cases of air pollution prevention financing funds provided to firms according to economic support plans, as well as the amount of these funds.

When considering personnel, not only the number of people, but also the quality of their work is important (Honadle, 2001). For empirical analysis of human capital in economic growth, Barro (1991) and Mulligan and Sala-i-Martin (1997) mainly use the number of years of education, labourer income and other representative variables as measurements of the quality of labour. Considering this, we set the average number of years employed as an indicator of experience and skill to measure the quality of personnel involved in environmental policies, and multiplied this by the number of staff to determine the overall value of personnel.

Analysis results

Analysis 1: identification of capacity elements

Through exploratory factor analysis (principle factor method, promax rotation) using the scree test as the standard, four factors were derived for both of the cities (Tables 3.2 and 3.3). For cronbach's α coefficient, Osaka's fourth factor was slightly low at 0.684, but the others were between 0.782 and 0.976, so we were able to confirm the existence of each factor with some confidence. The data structures of the factors were, with one exception, the same for both cities, making the same interpretation of factors for both cities possible.

For Kitakyushu, we interpreted the four factors as follows. Factor 1 was the 'environmental policy resource' management capacity provided by people, goods, money and other resources that support environmental policies. Factor 2 was the 'command and control' capacity indicated by onsite inspections of pollution emissions sources and specimen testing. Factor 3 was the capacity for 'financial support' to economically assist countermeasures by emitters of pollution. Factor 4 was the capacity to provide 'scientific knowledge' in order to create a basis for policy development and execution. For Osaka, on the other hand, factor 1 was the capacity for 'environmental policy resources' management, factor 2 was the capacity for 'financial support', factor 3 was the capacity to provide 'scientific knowledge', and factor 4 was the capacity to 'command and control'.

If we compare the data of the two cities for 'city/division environmental personnel', the data is the capacity to 'command and control' in Kitakyushu and the capacity to manage 'environmental policy resource' in Osaka. The factor ranks differ, but otherwise they have exactly

Table 3.2 Factor load and 4 factors (Kitakyushu)

Data	Factor 1	Factor 2	Factor 3	Factor 4
Budgets/municipal laboratory	0.933	−0.182	0.000	0.058
Sanitation expenses/city	0.819	−0.080	0.380	0.342
Personnel/municipal laboratories	0.733	0.310	0.411	0.347
Number of ambient air pollution monitoring stations	0.692	0.172	0.502	0.408
Environmental division personnel/city	0.096	0.915	−0.076	0.216
Number of onsite inspections of factories and workplaces related to air pollution/city	−0.229	0.855	0.167	0.024
Number of specimens analyzed for air pollution/municipal laboratories	0.133	0.707	−0.033	0.450
Amount of air pollution prevention financing funds provided to firms/city	−0.198	0.073	−0.818	−0.100
Number of cases of air pollution prevention financing funds provided to firms/city	−0.571	−0.372	−0.603	−0.286
Number of research presentations at academic conferences/municipal laboratories	0.394	0.253	0.170	0.864
Number of articles in academic journals/municipal laboratories	0.193	0.420	0.271	0.526
Eigenvalue	3.363	2.508	1.821	1.594
Contribution (%)	52.0	21.4	6.9	4.2
Cumulative contribution (%)	52.0	73.4	80.2	84.4

Table 3.3   Factor load and 4 factors (Osaka)

Data	Factor 1	Factor 2	Factor 3	Factor 4
Number of ambient air pollution monitoring stations	0.971	0.189	−0.034	0.002
Envionmental division personnel/city	0.832	0.443	0.220	0.216
Personnel/municipal laboratories	0.687	0.629	0.225	0.213
Budgets/municipal laboratory	0.665	0.604	0.257	0.317
Sanitation expenses/city	0.613	0.542	0.381	0.409
Amount of air pollution prevention financing funds provided to firms/city	−0.225	−0.952	−0.046	0.049
Number of cases of air pollution prevention financing funds provided to firms/city	−0.492	−0.827	−0.052	−0.108
Number of articles in academic journals/municipal laboratories	0.068	−0.029	0.992	−0.091
Number of research presentations at academic conferences/municipal laboratories	0.389	0.496	0.580	0.230
Number of onsite inspections of factories and workplaces related to pollution/city	0.020	−0.383	−0.503	−0.568
Number of specimens analyzed for air pollution/municipal laboratoies	−0.212	−0.103	0.043	−0.489
Eigenvalue	3.538	3.479	1.995	0.891
Contribution (%)	42.6	30.3	10.7	4.9
Cumulative contribution (%)	42.6	72.9	83.5	88.4

the same factor structures. The differences between 'city/division environmental personnel' results can be understood by explaining the air pollution policies of the two cities. In Kitakyushu, a main function performed by personnel is onsite inspection of a small number of firms based on pollution control agreements, while in Osaka a large number of firms are the subjects of administrative guidance. Therefore, the functions of government personnel in Osaka are wide-ranging and include the implementation of numerous explanatory meetings and technology investigations.

In addition, considering the contribution rate for each factor, factor 1, 'environmental policy resource' management, accounts for 52.0 per cent for Kitakyushu, and 42.6 per cent for Osaka City, showing that this accounts for roughly half of their capacities for air quality management. If we add factor 2 for each city, we can explain a little over 70 per cent of the capacity. Factor 2 was 'command and control' for Kitakyushu and 'financial support' for Osaka. This can be explained by their different policy measures based on their distinct urban structures. In Kitakyushu, where the subjects were a small number of large firms with significant technological and financial capabilities for environmental measures, the main activity of personnel is after-the-fact monitoring of firm countermeasures. In Osaka, where the subjects were a large number of medium and small firms without significant technological and financial capabilities for environmental measures, the main activity of city personnel is guidance and support of the firms' measures. This shows that a large part of the capacities of both cities can be explained by elements of capacity for the execution of policies suited to each city's characteristics, and by the elements of capacity for management of environmental policy resources that make policy development possible.

Regardless of their distinct urban structures and their use of differing policy measures, if we consider their cumulative contribution rates, this shows empirically that government capacities for air quality management in both cities can be explained by these four elements. Moreover, given the large contribution rates of factors 1 and 2 to elements of capacity, we confirmed the practicality and suitability of the air pollution policies of both cities.

Analysis 2: verification of capacity elements for air quality improvement

Making SO_2 concentration the dependent variable, and the factor scores of the four elements the explanatory variables, we conducted regression analysis using OLS. All coefficients for both cities had negative signs and were statistically significant, making clear the contribution of all four elements to SO_2 concentration decrease (Table 3.4). In addition, the standard partial regression coefficient shows that the 'environmental policy resource' management capacity of both cities was the highest contributor. Next in order for both cities were 'financial support', 'command and control', and 'scientific knowledge' provision.

Next, for the factor scores of each of the four elements, we set the contribution rate of their factor loads as their weights, calculated their weighted average values, and set these as their capacities for air quality management. Then, we conducted regression analysis using OLS with these as explanatory variables and SO_2 concentration as the dependent variable (Table 3.5). As a result, the coefficients for both cities were negative and statistically significant, confirming that capacities for air quality management contribute to air quality improvement.

In addition, we graphed the development process for capacities for air quality management from 1970 to 2000 (Figures 3.3 and 3.4). The development of capacities for air quality management for both cities can

Table 3.4   Regression analysis of 4 capacity elements

	Kitakyushu	Osaka
Scientific knowledge	−0.366 **	−0.113 *
	(−6.49)	(−2.51)
Environmental policy resource	−0.567 **	−0.881 **
	(−10.01)	(−19.55)
Command and control	−0.368 **	−0.222 **
	(−6.54)	(4.94)
Financial support	−0.501 **	−0.323 **
	(−8.85)	(−7.21)
Constant	0.010 **	0.016
	(18.06)	(26.79)
F value	72.587 **	118.367 **
Adj. R2	0.905	0.940

Note: ** $p < 0.01$, * $p < 0.05$.

Table 3.5   Regression analysis of capacity for air quality management

	Kitakyushu	Osaka
Capacities for air quality management	−0.807 **	−0.935 **
	(−7.36)	(−14.19)
Constant	0.010 **	0.016 **
	(9.26)	(18.19)
F value	54.155 **	201.247 **
Adj. R2	0.639	0.870

Note: ** p < 0.01.

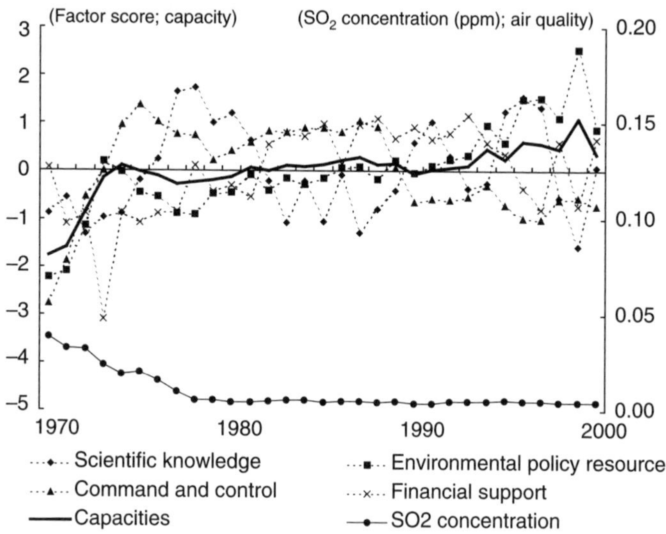

Figure 3.3   Trend of government capacities for air quality management (Kitakyushu)

be interpreted as growing rapidly from the early 1970s. Harashima and Morita (1995) classify the stages of Japan's environmental policy development as a progress period – when environmental policies were enhanced and strengthened and their results were recognized (1965–74), and a modification, stagnation and global environmental period – when the environmental problems that are the subjects of environmental policies changed qualitatively (1975–present). In comparison, Teranishi (1994) classifies the historical transformation of Japan's environmental policies

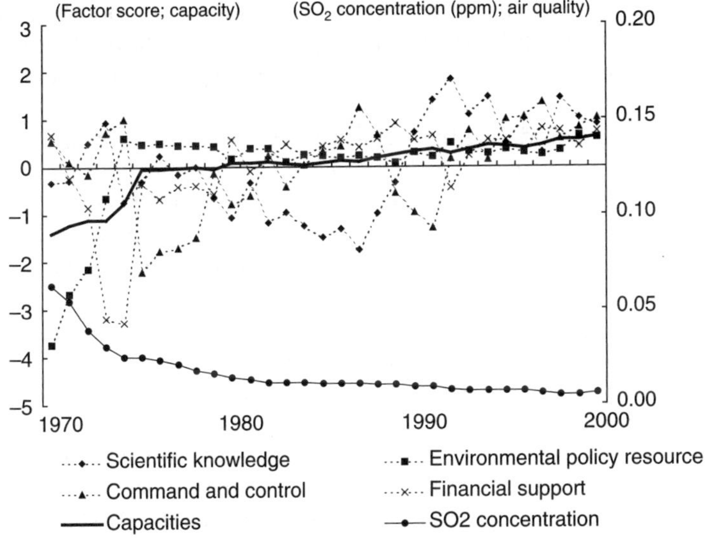

Figure 3.4   Trend of government capacities for air quality management (Osaka)

as the formation and development period (1965–74), the rapid change and retreat period (1975–88), and the muddling and reshuffling period (1989–present).

The formation processes of capacities for air quality management in this analysis are consistent with these environmental policy period classifications, and they also contribute to the policy results of air quality improvement. This shows that capacity, which is responsible for the efficacy of environmental policies, also developed in the same sequence and raised policy results. By confirming the relationship between air quality – which is the result of policy – and capacity – which is responsible for the efficacy of policy – we verified the validity of capacities for air quality management. In other words, we confirmed the existence of the four elements of capacity and their weights.

Setting assessment standards for capacities related to air pollution countermeasures

Towards capacity assessment standards for capacity elements

Through exploratory factor analysis, we identified four capacity elements and empirically confirmed that all of them have the function of contributing to the policy result of air pollution improvement. We also confirmed that the weight of each element was an appropriate reflection of its total contribution to capacity. This result indicates that these four elements are suitable as the elements of capacity that should be assessed.

Given this, considering that capacity levels are responsible for policy efficacy and the extent of policy results, by showing the roles of the four elements during policy cycles, we created assessment standards that link elements of capacity with policy cycles. By presenting the relationships between capacity elements according to their roles in the policy cycles, we can show capacity assessment standards in all stages of the policy cycle. This results in a capacity assessment framework that allows a diversified perspective during assessment.

In addition, policy cycles can be viewed from the perspective of knowledge management (Yamauchi *et al.*, 2000), and by expressing policy cycles as knowledge cycles, we made this a guide to setting the roles of elements of capacity in policy cycles. The correspondence of capacity elements with policy cycles and knowledge cycles not only clarifies the relationships between capacity elements, it also helps embody the capacity concept and has the practical merit of making data collection and other tasks easier during capacity assessment. Of course, since capacity is not something that has only specific purposes, the policy cycle stages when elements of capacity are effective cannot always be clearly divided, and some overlap is possible.

Proposal of a frame for capacity assessment of government air pollution countermeasures

The four elements of capacity fill different roles in policy cycles, as shown in Figure 3.5. The capacity to provide 'Scientific knowledge' fills roles of knowledge creation and accumulation in policy issue setting. The capacities of 'command and control' and 'financial support' fill roles in putting knowledge into practice in policy execution. The capacity for 'environmental policy resource' management fills both a knowledge use role in policy development and a knowledge reproduction role in policy assessment. In addition, as we will describe later, the capacity for 'environmental policy resource' management also influences the effectiveness of policy issue setting and policy execution.

'Scientific knowledge' fills the role of identifying issues that should be handled by policies with a scientific basis created through analysis of air pollution conditions, assessment and consideration of pollution elimination technology, investigative research, test inspections, and information gathering. According to Katsuhara (2001), the Kitakyushu

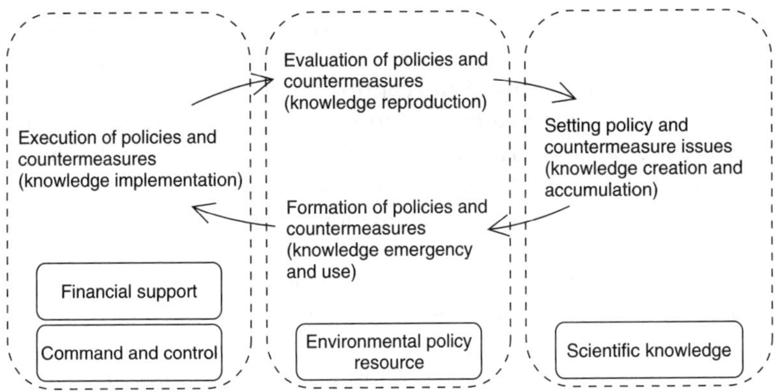

Figure 3.5   Correspondence of capacity elements with policy and knowledge cycles

City Institute of Environmental Sciences provides scientific data and information and makes the implementation of persuasive environmental governance possible. Fujikura (2002) states that governments are able to show the validity of scientific data when imposing pollution countermeasures on firms, and evaluates highly the early establishment of laboratories by both cities.

'Command and control' and 'financial support' also have roles in policy execution. 'Command and control' is the implementation of onsite inspections related to air pollution at factories and workplaces based on air pollution prevention laws, pollution prevention regulations, citizen complaints and other reasons. In addition to inspections of fuel and raw material samples, verifications of report details on equipment efficiency evaluations, maintenance and management conditions, standard compliant conditions, and inspections of the status of pollution prevention systems and the activities of pollution prevention managers and similar items are also conducted. They are also important in relation to the capacity to promote emission source countermeasures in response to violations, and include the issuing of improvement orders, improvement recommendations, improvement instructions and other administrative guidance.

'Financial support' is the provision of economic support for the pollution countermeasures of mainly medium and small firms. This includes the contribution of funds and loan assistance for the purchase of necessary machinery, the installation of equipment and the transfer of offices and facilities with the goal of pollution prevention. These systems are set in consideration of the difficulty faced by medium and small firms in meeting regulation standards and are rooted in an interdependent relationship with 'command and control'. As one aspect of air pollution policies, they fill the role of the carrot to the stick of 'command and control'. The capacity to implement 'financial support' is not simply the assurance of a financial framework, but it also includes the implementation of explanatory meetings, examinations, management oversight and other process capacities.

'Environmental policy resource' management includes obtaining facilities, equipment, personnel, budget and other resources, their appropriate distribution, and their efficient use at every stage of the policy cycle. Facilities and equipment include ambient air pollution monitoring stations, which provide information about both the need for environmental policies and their effectiveness. In short, they are resources for understanding the air pollution situation that are particularly important for policy development and policy assessment.

In addition, personnel are responsible for all actions related to environmental policies, and budgets are the frameworks that determine the extents of those environmental policies. Both personnel and budgets, as the means for carrying out political decision-making regarding what types of policies will be implemented and to what extent, fill important roles in policy development and policy assessment functions in particular. The reason for this is that, at the stages of policy issue setting and policy execution, purposes and the extent of application have been made clear to some degree, so the need for political decision-making can be said to be relatively low.

Furthermore, according to policy evaluation theory, we can explain the placement of policy development and policy assessment in the same types of roles. According to recent policy evaluation theory, policy development should occur after policy outcome goals are determined and policy assessment standards are set (Ueyama, 1998; Mayama, 2001). The reason for this is that, in policy cycles that spiral up, the results of developed policies are assessed and then they are reinvestigated. Thus, policy development and policy assessment are two aspects of the same policy process, and the capacity that supports their efficacy can be taken as the same. Therefore, the capacity for 'environmental policy resource' management, which allows the realization of political decision-making, can be explained as filling both policy development and policy assessment roles.

Thus, by showing the policy cycle roles of elements of capacity that we identified through empirical analysis, we are able to use these elements as capacity assessment standards. Given this, we can show the assessment frame for the capacity for government air pollution countermeasures as in Table 3.6.

Developing a capacity assessment framework

Background and purpose

We used exploratory factor analysis (EFA) to identify systematically the elements that form government air pollution countermeasure capacities, and we empirically confirmed the contribution of these elements of capacity to air quality improvement, resulting in standards for capacity assessment. However, this analysis was limited to the capacities of governments and disregarded the capacity of firms and citizens, which are also relevant social actors. Still, the government classifications shown in Table 3.6 for policy countermeasures and knowledge cycles are, to some extent, classifications based on ordinary standards, so they are also applicable in determining the capacity of firms and citizens.

In this section, we apply the government capacity assessment framework in Table 3.6 to determine empirically the capacity related to air pollution countermeasures of firms and citizens in addition to governments to clarify their roles in the formation of the social capacity for environmental management (SCEM) that contributes to air quality improvement. Through this, we are able to develop a capacity assessment framework that can evaluate SCEM and each actor's environmental management capacity level from observed data.

The rest of this section deals with analysis subjects and methods, and analysis results.

Subjects and methods of analysis

Setting analysis subjects

We chose two air pollution substances with different emissions sources and degrees of solution difficulty as the subjects of our analysis. In addition to sulphur dioxide (SO_2), which is a typical industrial air pollution caused by factories, workplaces and other fixed emissions sources already covered, we looked at nitrogen dioxide (NO_2), a typical substance emitted from not only fixed emission sources but also automobiles and other mobile emissions sources that contributes significantly to air pollution in the urban living environment.

We chose the time from when environmental policies were actually implemented and results began to be apparent for the analysis period. In this case, we chose the 30-year period from 1971–2000.

Table 3.6 Government capacity assessment framework (air quality management)

	Factor	
Process Execution of policies and countermeasures (knowledge implementation)	Evaluation of policies and countermeasures (knowledge emergence and use) Formation of policies and countermeasures (knowledge reproduction)	Setting policy and countermeasure issues (knowledge creation and accumulation)
Capacity P: Policy and countermeasure execution capacities (policy and measure)	R: Environmental policy and countermeasure resource management capacities (resource management)	K: Knowledge, information and technology provision capacities (knowledge and technology)
Actor (G = Government)		

• Regulatory methods
  • Creation and application of air pollution countermeasure legal regulations
  • Setting and enforcement of environmental and emission standards
  • Creation and application of air pollution
	• Funds, budgets
	  • Expansion of air pollution countermeasure budgets
	• Personnel and organization
	  • Establishment of environment related divisions
		• Investigation and research
		  • Research on air pollution causes, mechanisms, etc.
		  • Development of air pollution countermeasure technology and know-how accumulation
		  • Policy research on air pollution countermeasures

countermeasure ordinances and
fundamental plans
  - Observation through onsite
    inspection, etc.
- Market-based methods
  - Formation and application of
    environmental taxes, charges and
    subsidy payment systems
- Autonomous methods
  - Formation of pollution
    prevention agreements (without
    legal basis)
  - Promotion of environmental
    study and education

- Establishment of air pollution
  countermeasure organizations
  (committees, councils, business
  and citizen assemblies)
- Increase of air pollution
  countermeasure division staff
- Facilities and equipment, etc.
  - Arrangement of air pollution
    monitoring systems
  - Installation of air pollution
    warning equipment and
    arrangement of information
    systems

- Information disclosure and sharing
  - Disclosure of air quality
    conditions and pollution
    countermeasure information
  - Implementation of staff
    education and training

For our subject cities, we chose Osaka City and Kitakyushu City because they have different urban structures and they are both ordinance designated cities that have specified authority over environmental policies. These two cities have differences in fixed emission sources as shown in Figure 3.2. In addition, for mobile emission sources, the numbers of automobiles, the average traffic volumes and the average congestion levels (ordinary road total) are shown in Table 3.7. As a result, differences in annual average NO_2 concentration values in long-term continuous monitoring of automobile emissions occurred as shown in Figure 3.6. Furthermore, in Osaka Prefecture, a government designated area where the total NO_2 emissions volume is regulated that includes Osaka City, the automobile emissions ratio of the total NO_2 emissions amount was, according to estimates of the Japanese Ministry of the Environment, 47 per cent in 1985 (Japan Society of Air Pollution, 1993) and 51 per cent in 1997 (Osaka Prefectural Government, 2003). In contrast, in Kitakyushu City, the rate in 1983 was only 16 per cent (Kitakyushu City, 1998). From this, we can see that these two cities with their different urban structures also have different SO_2 and NO_2 emissions structures and countermeasures (priorities) (Figure 3.7).

With Osaka City and Kitakyushu City and their differing urban structures and air pollution countermeasures as our subjects, we investigated their structures of SCEM related to air pollution countermeasures and the influences on air quality improvement. Then, through comparison, we developed conclusions specific to each city and common to both.

Table 3.7 Numbers of automobiles, average traffic volumes and average congestion levels

	Number of automobiles		DID · average traffic volumes		DID · average congestion levels	
	Osaka	Kitakyushu	Osaka	Kitakyushu	Osaka	Kitakyushu
1971	603,547	176,515	29,951	17,058	0.95	0.81
1980	718,755	307,867	24,981	17,222	1.04	1.00
1990	939,728	447,254	25,620	18,880	1.36	1.11
1999	913,285	552,573	23,799	19,578	1.07	1.09

Notes: DID · densely inhabited district;
Average traffic volumes = sum (traffic × road length)/total road length;
Average congestion levels = traffic/traffic capacity.

Source: Road Bureau, Ministry of Land, Infrastructure and Transport (1971–99).

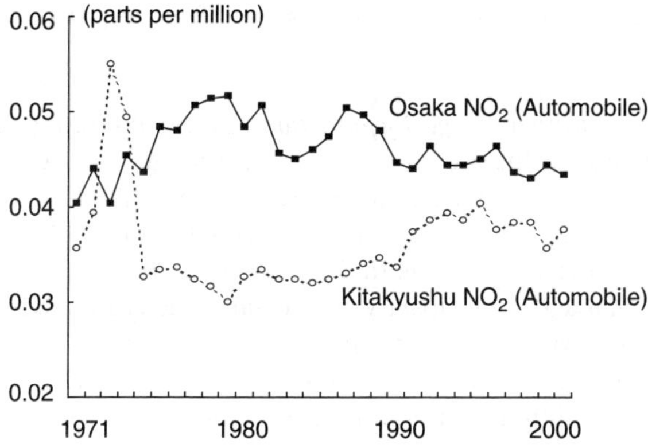

Figure 3.6   NO$_2$ concentration (automobile)
Source: Environmental Management Bureau, Ministry of Environment (1971–2000).

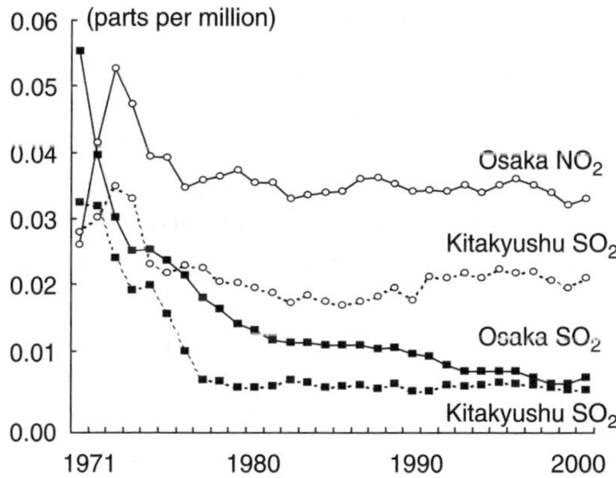

Figure 3.7   SO$_2$ and NO$_2$ concentration
Source: Environmental Management Bureau, Ministry of Environment (1971–2000).

Analytical methods

We sought to clarify the structures for social capacity for environmental management expressed by the environmental management capacities of governments, firms and citizens. Then, we made a formula for the causal relationship between SCEM and air quality improvement, and through verification, could show the structures of SCEM for air quality improvement. We conducted a two-stage empirical analysis to achieve this.

Analysis 1: verification of capacity formation   Through multi-group simultaneous analysis using a confirmatory factor analysis (CFA) model, we clarified the SCEM structures. Factor analysis is a method for identifying latent variables that cannot be directly observed from the observed series of variables (Yanai *et al.*, 1990). In this case, we derived each actor's environmental management capacity and social capacity for environmental management – which are latent and cannot be directly observed – from each actor's observable activity level in relation to air pollution countermeasures. In addition, multi-group simultaneous analysis is a method that investigates whether similar factor structures can be assumed across multiple populations. In this case, for both Osaka and Kitakyushu, to verify configural invariance, we assumed equivalent structures for the distribution of observed and latent variables (consistency of path position) and investigated both populations simultaneously to show the validity of the hypothesized capacity formation.

Analysis 2: verification of the causal relationship between capacity and air quality   We used a structural equation model (SEM),[2] which is a statistical approach that introduces latent variables that cannot be directly observed and identifies the causal relationship between these latent variables and observed variables in order to understand social and natural phenomenon (Kano and Miura, 2002). In this case, we formulated the social capacity for environmental management estimated in analysis 1 and the causal relationship with air quality improvement for each city. Doing this, we were able to verify the contributions of social capacity for environmental management to air quality improvement.

Data

As shown in Table 3.6, each actor's environmental management capacity can be expressed using three factors that correspond to the policy-countermeasure cycle and knowledge cycle classifications. From these, we selected data according to the three factors classifications for the series of phenomena related to the air pollution countermeasures of each actor's fundamental social role. We assumed these roles to be policies for governments, production for firms and consumption for citizens. In short, the relationship between each actor's environmental management capacity and the 'outcomes' of SO_2 and NO_2 concentration improvements can be expressed by the three data for 'input' (K, R) and 'output' (P) that indicate the levels of air pollution countermeasure achievements (Table 3.8).

Moreover, each actor's data were processed by different units that we then unified to create indexes. For governments, this was per city and environmental institute, for firms it was per workplace and for citizens it was per household.

Government environmental management capacity (G_cap) For governments, appropriate implementation of air pollution policies (G_P) can be set as the output of policies of air pollution countermeasures. Furthermore, behind these achievements is the scientific research (G_K) that is the basis for the policies and the organizational structures involved in making policies from that scientific research (G_R). Thus, a policy process series of G_K to G_R to G_P exists for air pollution countermeasures. In this case, the elements of capacity that correspond to these can be represented by the three items under G_cap in Table 3.8.

Regarding environmental governance, Matsushita (2002) has expressed the need to divide the policy process into the setting of issues, policy development and policy implementation, and has noted that there are gaps between the reality of environmental problems, the comprehension of them as governmental issues, policy setting and actual policy implementation. This indicates a need for investigation by policy process classification in order to grasp the causes of these gaps. Thus, setting scientific research, the human resources related to policy creation and policy implementation as individual elements of capacity that correspond to policy process classifications is appropriate.

Firm environmental management capacity (F_cap) For firms, we can set resource efficient production (F_P) as the output of air pollution countermeasures. Behind this achievement are the accumulation of technology and expertise related to resource efficient production (F_K) through cleaner production technology and the organizational structures for pollution prevention and production process management (F_R). Thus, the production process series related to air pollution can be described as F_K to F_R to F_P. In this case, the elements that correspond to these are represented in Table 3.8 by the three items under F_cap. In this table, we applied a GDP deflator to realize actual values of F_K. Moreover, tangible fixed assets at year-end (F_K) are not limited just to cleaner production technology, but are assets related to overall production activity. We can use this, though, because we can assume that it is rational behaviour

Table 3.8 Data used at actor analysis

		Data	Source
G_cap (Government)	G_K	Number of academic articles about research results from city environmental science laboratories	Report by Kitakyushu City Institute of Environmental Sciences
	G_R	Number of city environmental division and environmental science laboratory staff members × average number of years employed	Annual report of Osaka City Institute of Public Health and Environmental Sciences
	G_P	Number of air pollution inspections of factories, workplaces and other facilities by city divisions and environmental science laboratories	Environment in Kitakyushu City White Paper on Environment Pollution in Osaka City Survey on wages of Local Government Employees/Ministry of Internal Affairs and Communications
F_cap (Firms)	F_K	Actual value of manufacturing industry tangible fixed assets at the end of the year	Census of Manufacturers/Ministry of Economy, Trade and Industry
	F_R	Total number of executives and managers dedicated to (air) pollution prevention who have passed national tests	Environmental Management/Japan Environmental Management Association for Industry
	F_P	Value of products shipped by manufacturing industry/raw materials value	Annual report on National Accounts/Cabinet Office
C_cap (Citizens)	C_K	Actual expenditures on communications, publications and other printed matter	Annual report of family income and expenditure survey/Statistics Bureau
	C_R	Actual city social education expenses	Annual report on local finances/Ministry of Home Affairs
	C_P	Actual expenditures on buses and train passes for commuting to work and school	Base Linked Consumer Price Index Time Series/Statistics Bureau Annual report on national accounts/Cabinet Office
ENV. (Environment)	SO_2	SO_2 concentration (annual average values under ambient air pollution monitoring station)	Air Pollution in Japan/Ministry of Environment
	NO_2	NO_2 concentration (annual average values under ambient air pollution monitoring station)	

for firms to introduce equipment and facilities for resource efficient production and to accumulate technology and expertise for their operation. In addition, since pollution prevention managers and others who have passed national tests are tested at nine different sites throughout Japan, we calculated their number as: (total number of people who passed the test at each site) × (applicable urban population/population in jurisdiction of the applicable Regional Bureau of Economy, Trade and Industry) (Honda, 2004). Then, since the average age of the people who passed the first test in 1971 was about 30, we assumed that these personnel and their knowledge and expertise regarding pollution prevention have made continuous contributions, so we treated these years as accumulated data.

Regarding capacity base administration, Konno and Nonaka (1995) have classified that capacity in three layers: the organizational resources level (knowledge resources, cognitive capacity), the knowledge conversion level (systems, processes, skills) and the product level (core products, complementary products and services). They state that the interrelations between these result in a firm's total capacity. This corresponds to the items in this research expressed as a firm's K, R and P.

Citizen environmental management capacity (C_cap)   We set the active use of public transportation (C_P) as the citizen consumption output of air pollution countermeasures. Behind the realization of this behaviour is knowledge and information that cultivate environmental consciousness (C_K) and the creation of situations that promote individual consciousness toward environmental behaviour (C_R). The consumption process series related to air pollution can be seen as C_K to C_R to C_P. The elements of capacity that correspond to these are shown in Table 3.8 by the three items under C_cap. Moreover, we adjusted C_K and C_P for each city and item using deflators. We used education expenditures as the deflator for C_R.

Stern (2000) has summarized research on individual environmental behaviour and the primary factors that influence it, and has classified environmental behaviour as '1 environmental activism', '2 non-activist behaviours in the public sphere', '3 private-sphere environmentalism', and '4 other.'

The primary factors of these environmental behaviours could be classified as '(1) attitudinal', '(2) personal capabilities', '(3) external/contextual forces', '(4) habit and routine', and it could be recognized that specific environmental behaviour is induced through the mutual interaction of multiple primary factors.

In this analysis, we used environmental public transportation system use as behaviour 3 (private-sphere environmentalism (C_P)). For its primary factors, we set (1) attitudinal and (2) personal capabilities as 'knowledge and information that cultivate environmental consciousness' (C_K) and (3) external/contextual forces and (4) habit and routine as 'the creation of situations that promote individual consciousness toward environmental behaviour' (C_R).

'Actual expenditures on communications, publications and other printed matter' (C_K) represent thoughts, senses of value and fundamental capacity levels regarding the environment. Next, we looked at personal and mass communication classifications, which are information circulation structure classifications related to the creation of Hiromatsu's (1986) information index. Telephones, postcards and similar communication expenses were counted as personal communications, while newspapers, magazines and other publications and printed matter expenses were considered mass communication, and these together formed our information and knowledge index.

'Actual city social education expenses' (C_R) includes public meeting place expenses, library expenses and social education activity expenses, and contributes to social standards and the maintenance of customs and manners in the public sphere. In addition, externality of education is also a factor. In short, if decisions about education needs are entrusted to individuals, there is a chance that the desired education level of society as a whole may not be achieved (Oshio, 2002), so education instituted by society complements individual knowledge and information, and encourages activity as a group. Regarding these, knowledge originates amid specific relationships between times, places and other people (Nonaka *et al.*, 2003), so shared education can be said to contribute to the creation of these opportunities. In addition, education by society supports the activities of groups involved in such education, including protection of the natural environmental and acting as watchdogs for environmental problems.

For this analysis, we set SO_2 concentration and NO_2 concentration as the annual average values from ambient air pollution monitoring station.[3]

Analysis results

Analysis 1: verification of capacity formation

Using the confirmatory factor analysis model in Figure 3.8, we conducted multi-group simultaneous analysis with results as shown in Table 3.9.

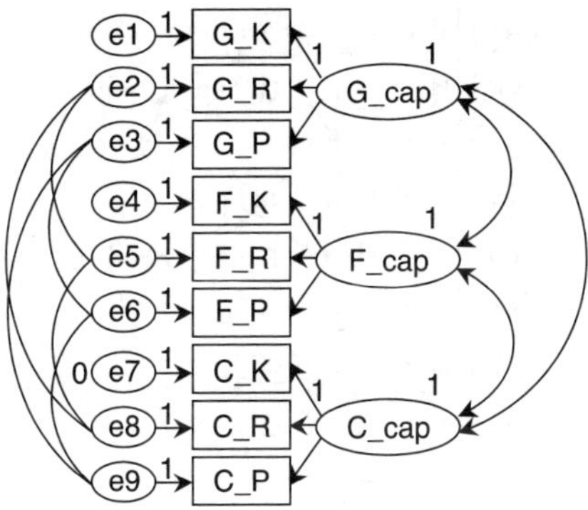

Figure 3.8   Primary factor analysis model

Table 3.9   Results of multi-group simultaneous analysis

			Osaka	Kitakyushu
G_cap	↔	F_cap	0.88 ***	0.76 ***
G_cap	↔	C_cap	0.94 ***	0.78 ***
F_cap	↔	C_cap	0.94 ***	0.95 ***
	→	G_K	0.90 ***	0.73 ***
G_cap	→	G_R	0.77 ***	0.87 ***
	→	G_P	0.94 ***	0.85 ***
	→	F_K	1.00 ***	0.69 ***
F_cap	→	F_R	0.90 ***	0.99 ***
	→	F_P	0.93 ***	0.82 ***
	→	C_K	1.00 ***	1.00 ***
C_cap	→	C_R	0.38 *	0.72 ***
	→	C_P	0.70 ***	0.29

$\times 2 = 50.399$ (df 39), P = 0.104, GFI = 0.807

Note: *** p < 0.01, *p < 0.10.

Since the correlation between each actor's capacity for environmental management is high, we hypothesized the existence of shared higher latent concepts as the social capacity for environmental management and created the secondary factor analysis model in Figure 3.9, with results as shown in Table 3.10.

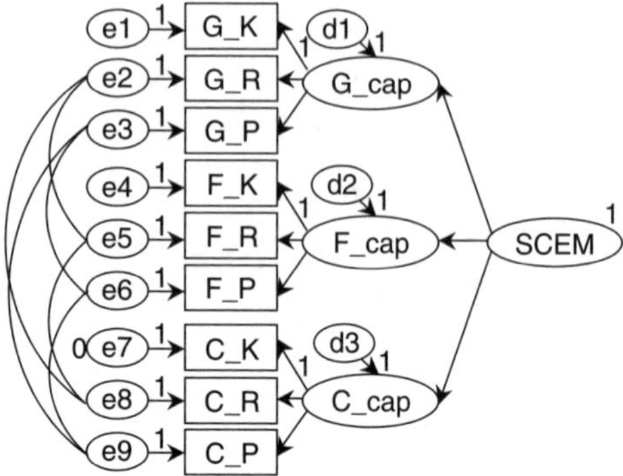

Figure 3.9   Secondary factor analysis model

Table 3.10   Results of secondary factor analysis model

			Osaka	Kitakyushu
	→	G_cap	0.94 ***	0.79 ***
SCEM	→	F_cap	0.93 ***	0.96 ***
	→	C_cap	1.00 ***	0.99 ***
	→	G_K	0.90 –	0.73 –
G_cap	→	G_R	0.77 ***	0.87 ***
	→	G_P	0.94 ***	0.85 ***
	→	F_K	1.00 –	0.69 –
F_cap	→	F_R	0.90 ***	0.99 ***
	→	F_P	0.93 ***	0.82 ***
	→	C_K	1.00 –	1.00 –
C_cap	→	C_R	0.38 *	0.72 ***
	→	C_P	0.70 ***	0.29

$\times 2 = 50.399$(df 39), P $= 0.104$, GFI $= 0.807$

Notes: ***p < 0.01, * p < 0.10.
Since the paths to K are set as 1 in order to achieve model
discrimination, CR (critical ratio) is not calculated.

In each model, we set the covariance between each actor's R and P error
variables.[4] In addition, to assure discrimination for each model,[5] for the
primary factor model in Figure 3.8, we fixed each actor's environmental
management capacity variance as 1. For the secondary factor model in
Figure 3.9 we sought a solution by setting SCEM variance as 1, and the

latent variables to K paths coefficient was set as 1. In addition, judging them to be sample variances (Kano and Miura, 2002; Toyoda, 2003), we set both cities' e7 error variables and Osaka's e4 error variable variances as 0.[6]

We calculated model fitness jointly for both populations and found $\chi^2 = 50.399$ (df 39), P $= 0.104$ (>0.05), and GFI $= 0.807$ for both models, indicating a degree of fitness.[7] Moreover, GFI is somewhat low, but this is the result of the small sample size (Kano, 1998).[8] All of the standardized path coefficients are positive, high values, so we can say that the latent variables were suitably measured by the observed variables, and the ability to explain cause and effect is high. In addition, except for Kitakyushu's C_cap to C_P, every path coefficient was statistically significant. From this, we can say that structures of social capacity for environmental management (SCEM) for both cities were equally expressed. Moreover, Kitakyushu's C_cap to C_P path that was not statistically significant was reconfirmed in analysis 2.

At this point, interpretation of the model in Figure 3.9, which is formed from regression models, can be expressed, for example, by the following equation for the top SCEM to G_cap to G_K causal relationship for Osaka:

$$G\_cap = 0.94 \times (SCEM) + d1 \tag{3.1}$$

$$G\_K = 0.90 \times (G\_cap) + e1 \tag{3.2}$$

(d1: disturbance variable, e1: error variable)

This shows that the high level of social capacity for environmental management is related to the level of government environmental management capacity being high, and also the level of government capacity to provide knowledge, information and technology (K) being high. Thus, we show that the level of latent SCEM can be estimated from observed data.

Analysis 2: verification of the causal relationship between capacities and air quality

Using the structural equation model, the causal relationship between the social capacity for environmental management estimated using the secondary factor model of analysis 1 and the air quality improvement is as shown in Figure 3.10 and Figure 3.11. In order to achieve model discrimination, in each figure the highest path coefficient for the paths from latent variables (path to K, SCEM to G_cap, ENV to SO_2) was set as 1, and a solution was sought with this constraint. In addition, we

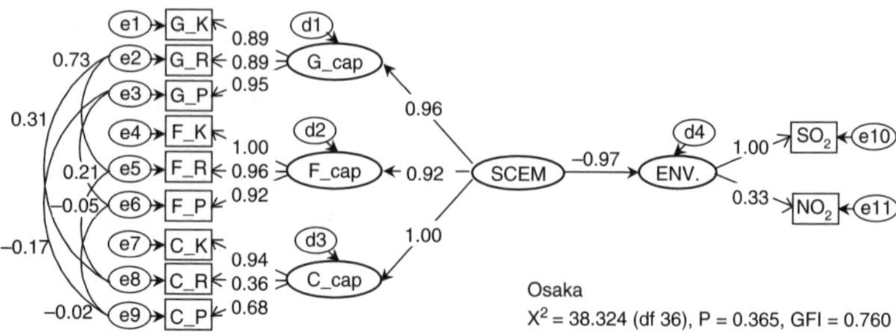

Figure 3.10   Causal relationship between capacities and air quality (Osaka)

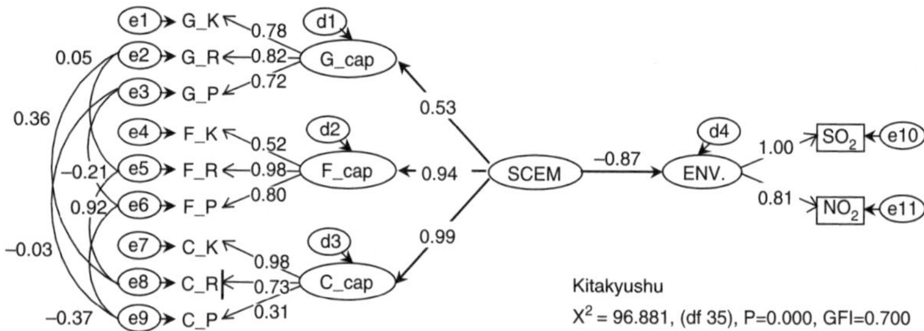

Figure 3.11   Causal relationship between capacities and air quality (Kitakyushu)

evaluated the influence of sample variance, and set the variance of the error variables e10 for both cities and e4 for Osaka as 0.

The model fitness for Osaka was $\chi^2 = 38.324$ (df 36), $P = 0.365$ (>0.05), and GFI = 0.760, while for Kitakyushu it was $\chi^2 = 96.881$ (df 35), $P = 0.000$ (<0.05), and GFI = 0.700. In addition, standardized path coefficient values were large. SCEM to ENV was negative, but the rest were all positive, showing that increasing the SCEM and its component elements has a causal relationship with the reduction of air pollution. In addition, all were significant at 5 per cent with the exceptions of Osaka's C_cap to C_R and ENV to NO$_2$, and Kitakyushu's C_cap to C_P.[9]

The Osaka model shows a certain degree of fitness, and the contribution of social capacity for environmental management to air quality improvement was verified.

On the other hand, the fitness of the Kitakyushu model was not especially good. One probable main reason for this is that there were many observed variables in this model and only a small number of samples (Toyoda, 2003).

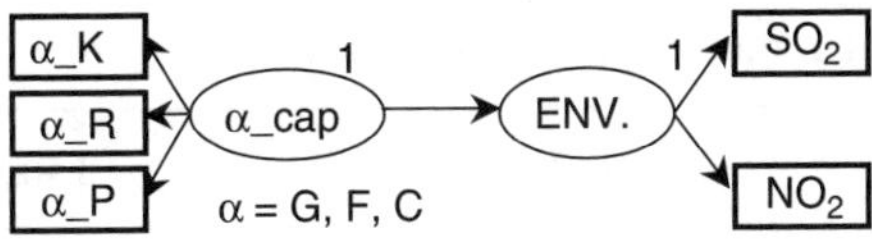

Figure 3.12   Model for each of the actors

Note: Error variable and disturbance variable are omitted.

For this reason, using a simpler model, we sought to confirm the causal relationship between the capacity for environmental management and air quality improvement for Kitakyushu, and to verify the paths that were not statistically significant at analyses 1 and 2. For this purpose, we used a structural equation model for each of the actors in Figure 3.12 to verify the causal relationship of each actor's environmental management capacity and air quality improvement.[10] Moreover, capacity assessment for each actor was conducted on actual assessment sites, and confirmation and consideration of causal relationships were conducted not only for the whole but also for the parts. To achieve model discrimination, G_cap (F_cap, C_cap), variance was set as 1 and the ENV to SO_2 path coefficient was set as 1, and solutions were sought with these constraints. In addition, we set SO_2 error variable variance as 0, judging it to be the influence of sample variance.

Table 3.11 shows that the fitness of all the models is good and that all the path coefficients are statistically significant. From this, we confirmed each actor's environmental management capacity formation and contributions to air quality improvement, complementing the results of analyses 1 and 2. In addition, through simultaneous analysis, a certain fitness was also achieved for distribution invariance, and the causal relationship structures of each actor's environmental management capacity and air quality improvement for both cities can be said to have been shown equally. Thus, we verified the contribution of social capacity for environmental management to air quality improvement.

Based on the above verification from observed data of the capacity formation of firms and citizens, in addition to governments, we have shown the applicability of the capacity assessment framework in Table 3.6 for an assessment standards system (index system) for estimating the level of SCEM for air quality improvement. From this, we created the Actor-Factor Matrix as a capacity assessment framework that can estimate the level of social capacity for environmental management (SCEM) (Table 3.12).

86

Table 3.11 Results of model for each of the actors

			Osaka	Kitakyushu
G_cap	→	G_K	0.90 ***	0.82 ***
	→	G_R	0.96 ***	0.91 ***
	→	G_P	0.72 ***	0.82 ***
	→	ENV.	−0.99 ***	0.73 ***
ENV.	→	SO_2	1.00	1.00
	→	NO_2	0.34 *	0.82 ***
Individual			$\times2 = 8.784$(df 5)	$\times2 = 8.784$(df 5)
			P = 0.118	P = 0.072
			GFI = 0.879	GFI = 0.860
Simultaneous			$\times2 = 18.912$(df 10), P = 0.041, GFI = 0.870	

			Osaka	Kitakyushu
F_cap	→	F_K	0.94 ***	0.53 **
	→	F_R	0.99 ***	0.95 ***
	→	F_P	0.98 ***	0.84 ***
	→	ENV.	−0.94 ***	0.93 ***
ENV.	→	SO_2	1.00	1.00
	→	NO_2	0.49 **	0.85 ***
Individual			$\times2 = 11.030$(df 5)	$\times2 = 10.069$(df 5)
			P = 0.051	P = 0.073
			GFI = 0.848	GFI = 0.861
Simultaneous			$\times2 = 21.100$(df 10), P = 0.020, GFI = 0.854	

			Osaka	Kitakyushu
C_cap	→	C_K	0.96 ***	1.00 **
	→	C_R	0.74 ***	0.72 ***
	→	C_P	0.73 ***	0.35 **
	→	ENV.	−0.98 ***	−0.84 ***
ENV.	→	SO_2	1.00	1.00
	→	NO_2	0.33	0.81 ***
Individual			$\times2 = 8.299$(df 5)	$\times2 = 8.203$(df 5)
			P = 0.141	P = 0.145
			GFI = 0.907	GFI = 0.909
Simultaneous			$\times2 = 16.501$(df 10), P = 0.086, GFI = 0.908	

Notes: *** p < 0.01, ** p < 0.05, * p < 0.10.
We set SO_2 error variable variance as 0, judging it to be the influence of sample variance.
Thus, ENV. → SO_2 path is to be 1.00.

Table 3.12 Actor–factor matrix (air quality management)

	Factor	
Process		
Execution of policies and countermeasures (knowledge implementation)	Evaluation of policies and countermeasures (knowledge emergence and use) Formation of policies and countermeasures (knowledge reproduction)	Setting policy and countermeasure issues (knowledge creation and accumulation)
Capacity		
P: Policy and countermeasure execution capacities (policy and measure)	R: Environmental policy and countermeasure resource management capacities (resource management)	K: Knowledge, information and technology provision capacities (knowledge and technology)
	Actor (G = Government)	

• Regulatory methods
  • Creation and application of air pollution countermeasure legal regulations
  • Setting and enforcement of environmental and emission standards
  • Creation and application of air pollution
	• Funds, budgets
	  • Expansion of air pollution countermeasure budgets
	• Personnel and organization
	  • Establishment of environment-related divisions
		• Investigation and research
		  • Research on air pollution causes, mechanisms etc.
		  • Development of air pollution countermeasure technology and know-how accumulation
		  • Policy research on air pollution countermeasures

Continued

Table 3.12 Continued

countermeasure ordinances and fundamental plans
- Observation through onsite inspection, etc.
- Market based methods
  - Formation and application of environmental taxes, charges and subsidy payment systems
- Autonomous methods
  - Formation of pollution prevention agreements (without legal basis)

- Establishment of air pollution countermeasure organizations (committees, councils, business and citizen assemblies)
- Increase of air pollution countermeasure division staff
- Facilities and equipment, etc.
  - Arrangement of air pollution monitoring systems
  - Installation of air pollution warning equipment and arrangement of information systems

- Information disclosure and sharing

  - Disclosure of air quality conditions and pollution countermeasure information
  - Implementation of staff education and training

Actor (F = Firms)

- Regulatory methods
  - Observance of air pollution countermeasure laws and regulations
  - Observance of environmental and emission standards
  - Compliance with air pollution countermeasure ordinances and fundamental plans

- Funds, budgets
  - Expansion of air pollution countermeasure budgets
- Personnel and organization
  - Establishment of environment related divisions
  - Increase of environment related division staff

- Investigation and research
  - In-house monitoring of factories and workplaces
  - Development of air pollution countermeasure technology and accumulation of expertise

- Market based methods
  - Countermeasures through the use of subsidy payment systems, etc.
- Autonomous methods
  - Formation of pollution prevention agreements (without legal basis)
  - Environmental load reduction through the entire business process including financial service production
  - Acquisition of ISO 14001 certifications, introduction of ESCO projects, etc.

- Increase numbers of environmental managers and pollution prevention managers
- Facilities and equipment, etc.
  - Arrangement of in-house monitoring systems
  - Arrangement of warning equipment and information systems

- Information disclosure and sharing
  - Creation and disclosure of environmental reports and environmental accounting
  - Implementation of staff education and training

Actor (C = Citizens)

- Regulatory methods
  - Observance of air pollution countermeasure laws and regulations (field burning, etc.)

- Funds, budgets
  - Expansion of environmental countermeasure budgets
- Personnel and organization

- Investigation and research
  - Investigation and research (NGO, NPO)
  - Observation and monitoring (NGO, NPO)

Continued

Table 3.12   Continued

• Market-based methods
  • Countermeasures that make use of subsidy payment systems, etc. (NGO, NPO)
• Autonomous methods
  • Complaints, requests and lobbying
  • Transition to lifestyles that conserve energy and resources, including more ecological automobile use, public transportation use, etc.
  • Green purchasing and environmental funds
	• Increase of the number of environmental countermeasure personnel (NGO, NPO)
	• Participation in NGO and NPO activities
	• Participation in environmental events, etc.
	• Facilities and equipment, etc.
	  • Acquisition of facilities and equipment related to environmental countermeasures (NGO, NPO)
	  • Introduction of environmentally friendly products (energy conservation, new energy devices)
		• Information disclosure and sharing
		  • Understanding air pollution status
		  • Implementation of environmental education and training

Advantage of the actor–factor matrix

The actor–factor matrix in Table 3.12 shows the core of the assessment details that we determined quantitatively with statistical data. We classified the factors according to their three roles as elements of capacity in the policy countermeasure and knowledge cycles. Combined with the three actor classifications, a 9-cell matrix is formed. In actual assessments, this matrix can be used to understand the capacities for air pollution countermeasures based on qualitative information from interviews and questionnaires in addition to statistical data.

This actor–factor matrix is a capacity assessment method that was developed after the elucidation of systemic elements of capacity, the confirmation of the contribution of these elements of capacity to air quality improvement, and the verification and setting of the assessment framework based on the policy countermeasure and knowledge cycles. Compared to the capacity assessment methods developed by the UNDP (1998) and other organizations with three classifications – individuals, organizations and social systems – this method can be said to be superior for the following three points:

Systematic structure of assessment standards

For social system capacity elements, the UNDP (1998) method gives policies, laws and regulations, management accountability, resources (people, budgets, information) and processes. Then, for organizations, it gives mission and strategy, organizational culture and structure, capacity, process, resources (people, budgets, information) and infrastructure. However, these do not clarify the necessary and sufficient conditions of the elements of capacity or their relationships in internal capacity structures, and they are not systematic assessment standards. For this reason, the UNDP framework is insufficient for capacity assessment data gathering and selection. In the actor–factor matrix, systemic capacity is derived through empirical analysis, and capacity elements are set in a form that corresponds to policy countermeasure and knowledge cycles. By making guidelines from coordinated classifications of policy countermeasure cycles, knowledge cycles and capacity elements, systematic gathering and selection of data related to capacity elements becomes easy.

Efficient and effective assessment implementation

The ambiguity and malleability of the UNDP (1998) assessment framework result in the creation of large checklists for each individual environmental problem. Furthermore, the relationships and roles in policy cycles of the capacity elements that are given to be checked are unclear, its uniqueness makes relationships with other assessment standards difficult to see, and as assessment standards they are hindered by repetitiveness. As a result, it is also impossible to confirm whether all checklist items are appropriate assessment standards for the contribution to policy results, or even if there is a need to check them all. The actor–factor matrix makes assessment standards from the elements of capacity that should be assessed because they have verified contributions to policy results, and it gives precedence to the efforts and effects related to the assessments. The actor–factor matrix is also suitable because it is a simple assessment method that developing countries can implement themselves, which is valuable for developing ownership and self-assessment capacity in those countries. Table 3.12 is an illustration of an air pollution countermeasure matrix that classifies capacity by the ordinary standards of policy countermeasure cycles, knowledge cycles and elements of capacity. Since the lower assessment items are also standard item classifications, this could also be used as a shared assessment framework for other environmental fields.

Appropriate policy proposals for capacity development

Since the UNDP (1998) methods treat relationships among individuals, organizations and social systems as a comprehensive layered relationship structure, it is impossible to determine where resources should be applied from the results of capacity assessments for individual layers. In contrast, the actor–factor matrix classifies three actors – governments, firms and citizens – and elements of capacity (and their places in the policy and knowledge cycles) to create nine cells with no overlap, making the division of roles clear. Thus, based on capacity assessment results, considering the possibilities of substitutions between actors, it is clear where resources should be applied in the matrix, making possible more effective policy proposals for capacity development.

However, social capacity for environmental management is dependent on the interrelationships between socioeconomic conditions and environmental quality, so in actual assessments, SCEM should not be the only focus, but rather total systems based on these interrelationships must also be considered.

To solve environmental problems that are becoming more diverse and complex, not only governments, but also firms and citizens need to fulfil appropriate roles. For this to be possible, though, each actor must have attained the capacity necessary to fill its role. Amid a tide of changing political systems – including democratization, devolution, privatization and private sector activation – effective, efficient solutions for social problems are needed not only for the environment, but also for other issues such as development assistance, government administration and community planning. These should be solved through the participation and cooperation of diverse actors. Capacity assessments for setting the division of roles for each actor and verification of efficacy are needed along with medium- and long-term capacity development plans that take these into account. We believe that the actor–factor matrix can be applied to these types of capacity assessments.

Notes

1   Civil society includes NGOs, NPOs and similar groups but, in this chapter, we focus on empirical analysis citizens as the subject. Furthermore, in our definition of social capacity for environmental management, we use the word 'citizen' as representative of the range of civil society.

2   Furthermore, the factor analysis in analysis 1 is also a subordinate model of SEM. In addition, structural equation models can also be called models for conducting factor analysis and regression analysis simultaneously. Moreover, we used Nonaka and Konno's (2003) concept creation methodology, which has steps as follows: (1) observation (idea prototype generation); (2) generalization (mechanism understanding); (3) modelling (causal relationship discovery); and (4) application (index and measuring of variables). A structural equation model is used as a method for (3) modeling and (4) application of the concept of social capacity for environmental management.

3   Ordinarily, pollutant concentration distributions in environmental media are expressed by a lognormal distribution, and even a histogram that is close in form to a lognormal distribution. For this reason, in order to use analysis methods that hypothesize a multivariate normal distribution as the population distribution, we made SO_2 and NO_2 concentrations logarithmic to approximate a normal distribution in our analysis.

4   In cases where there were no obvious reasons to assume otherwise, normally the covariance between error variables is assumed to be 0, but in cases when common variance factors other than latent variables were estimated to exist, we can assume covariance between error variables (Toyoda 2003). For this reason, the relatively easily transferable explicit knowledge that corresponds to R and P, in comparison with the implicit knowledge that corresponds to K indicated by Nonaka and Takeuchi (1996), has high relative fluidity, so we estimated that primary factors other than for those stipulated by each actor's capacity for environmental management also have an influence and we set covariances.

5   A situation where a model cannot be discriminated is when a single solution for the sought free parameter and the equation's numerical relationship in the assumed model cannot be determined. For this reason, in order to

achieve discrimination for the model, it is necessary to apply constraints to, for example, fix the path coefficient and variance. A common constraint is to set latent variable variance as 1 and a path coefficient from the latent variable to the observed variable as 1.

6 Error variance is always nonnegative. However, for an observed variable with very strong factor loads (path coefficients), the error variance becomes a small value and there is the possibility that it might even fall into a negative range. In such cases, error variance can be estimated and fixed at 0 (Toyoda, 2003).

7 In this χ^2 test, by making the null hypothesis be 'the model is correct' since the hypothesis will not be rejected with a P value greater than 0.05, a model with a 5 per cent significance level can be adopted.

8 The greater the sample number, the more reliable the results, but the necessary minimum sample number depends on the model, and there is no absolute standard (Toyoda, 2003). Toyoda (1998) suggests that a sample number of 30 can provide a stable solution in estimation of solutions for the specific concepts of specific individuals.

9 Among paths from a latent variable, the highest path coefficients (the path toward K, SCEM to G_cap, and ENV to SO_2) in the figure were set at 1 in order to achieve discrimination for the model. CR (Critical Ratio) were not calculated, so significance could not be evaluated, but analysis 1 resulted in judgments of significance.

10 From the primary factor model of analysis, the correlation with each actor's environmental management capacity was strong. Multicollinearity occurs when using the same model for verification of the relationship between the three actors' environmental management capacities and air quality improvement, so we verified using a simple model for each actor separately.

References

Barro, R.J. (1991) 'Economic Growth in a Cross Section of Countries', *Quarterly Journal of Economics*, 106: 407–43.

Boesen, J. and A. Lafontaine (1998) *The Planning and Monitoring of Capacity Development in Environment (CDE) Initiatives* (Canada: CIDA).

Environmental Management Bureau, Ministry of the Environment (1971–2000) *Air Pollution in Japan* (Tokyo) (in Japanese).

Fujikura, R. (2002) 'SO_x Control Measures of Local Government in Japan', in T. Terao, and K. Otsuka, *Dynamism of the Policy Process in Development and the Environment: the Experiences of Japan and Problems in East Asia* (Chiba: Institute of Developing Economies) (in Japanese): 37–78.

Harashima, Y. and T. Morita (1995) 'A Comparative Analysis of Environmental Policy Development Process in East Asian countries', *Planning Administration*, 18(3): 73–85 (in Japanese).

Hiromatsu, T. (1986) 'Information Technology Indicator and Information Census' in S. Hayashi and T. Nakamura: *Japanese Economy and Economic Statistics* (University of Tokyo Press) (in Japanese): 97–115.

Honadle, B.W. (2001) 'Theoretical and Practical Issues of Local Government Capacity in an Era of Devolution', *Journal of Regional Analysis and Policy*, 31 (1): 77–90.

Honda, N. (2004) 'The Role of the Social Capacity for Environmental Management in Air Pollution Control: An Application of Three Pollution Problems in Japan', *Papers on Environmental Information Science*, 18: 331–6 (in Japanese).

Janicke, M. and H. Weidner (eds) (1997) *National Environmental Policies: A Comparative Study of Capacity-building* (Berlin: Springer).

Japan Society of Air Pollution (1993) *Global Air Pollution and Measures* (Ohmsha) (in Japanese).

JICA (2004) *Handbook of Capacity Development: Improve Validity and Sustainability of JICA Activity* (Tokyo: JICA) (in Japanese).

Kano, Y. (1998) 'Analysis for Data of Favorable Impression to Entertainer', H. Toyoda (ed.) *Covariance Structure Analysis: A Case Study* (Kyoto: Ohmsha Shobo) (in Japanese): 9–21.

Kano, Y. and A. Miura (2002) *Graphical Modelling and Multivariate Statistical Analysis of AMOS, EQS and CALIS* (Kyoto: Gendai-Sugakusha) (in Japanese).

Katsuhara, T. (2001) *Development and Environmental Problem in East Asia* (Tokyo: Keiso Shobo) (in Japanese).

Kitakyushu City (1998) *Environmental Pollution Control in Kitakyushu City: analysis* (Kitakyushu City: Kitakyushu) (in Japanese).

Konno, N. and I. Nonaka (1995) *Intellectualizing Capability* (Tokyo: Nihon Keizai Shimbun), (in Japanese).

Lavergne, R. and J. Saxby (2001) *Capacity Development: Vision and Implications*, CIDA Policy Branch, Capacity Development Occasional Paper Series, 3 (Canada: CIDA).

Matsuoka, S. and Honda, N. (2002) 'Environmental Cooperation and Capacity Development: Review of the Concept of Capacity Development in Environment (CDE)', *Journal of International Development Studies*, 11 (2): 149–72 (in Japanese).

Matsuoka, S. and A. Kuchiki (eds) (2003) *Social Capacity Development for Environmental Management in Asia: Japan's Environmental Cooperation after the Johannesburg Summit* (Chiba: Institute of Developing Economies) (in Japanese).

Matsuoka, S., S. Okada, K. Kido, and N. Honda (2004) 'Development of Social Capacity for Environmental Management and Institutional Change', *Journal of International Development Studies*, 13 (2): 31–50 (in Japanese).

Matsushita, K. (2002) *Environmental Governance* (Tokyo: Iwanami Shoten), (in Japanese).

Mayama, T. (2001) *Essence of Policy Design* (Tokyo: Seibundoh), (in Japanese).

Mulligan, C.B., and X. Sala-i-Martin (1997) 'A Labour Income-based Measure of the Value of Human Capital: An Application to the States of the United States', *Japan and the World Economy*, 9 (2): 159–91.

Nonaka, I., H. Izumida, and A. Nagata (2003) *Towards the theory of a knowledge-based Country: A New Paradigm of the Policy Process* (Tokyo: Toyo Keizai) (in Japanese).

Nonaka, I. and N. Konno (2003) *Methodology of Knowledge Creation* (Tokyo: Toyo Keizai) (in Japanese).

Nonaka, I. and H. Takeuchi (1996) *The Knowledge-creating Company: How Japanese Companies Create the Dynamics of Innovation* (Tokyo: Toyo Keizai), (in Japanese).

OECD/DAC (1999) *Donor Support for Institutional Capacity Development in Environment: Lessons Learned. Evaluation and Effectiveness*, 3, (OECD), http://www.oecd.org/dataoecd/10/27/2667310.pdf (accessed at 19 August, 2005).

Osaka City (1994) *Environmental Pollution Control in Osaka City* (Osaka: Global Environment Centre Foundation) (in Japanese).

Osaka Prefectural Government (2003) *Total Emission Reduction of automobile Nitrogen Oxides and Particulate Matter in Osaka Prefecture* (Osaka Prefectural Government), (in Japanese).

Oshio, T. (2002) *The Economic Analysis of the Japanese Education* (Tokyo: Nippon-Hyoron-Sha) (in Japanese).

Research and Statistics Department, Economic and Industrial Policy Bureau, Ministry of Economy, Trade and Industry (1970–2000).

Research and Statistics Department, Economic and Industrial Policy Bureau, Ministry of Economy, Trade and Industry (1970–2000) *Census of Manufacturers* (Tokyo) (in Japanese).

Road Bureau, Ministry of Land, Infrastructure and Transport (1971–1999) *Road Traffic Census* (Tokyo) (in Japanese).

Stern, P.C. (2000) Toward a coherent theory of environmentally significant behaviour. *Journal of Social Issues*, 56 (3): 407–24.

Teranishi, S. (1994) 'A Critical Review of Pollution Issues and Environmental Policy in Japan: Lesson for NIEs', in R. Kojima and S. Fujisaki (eds), *Development and Environment: A Problem of Industrializing Asia*, (Institute of Developing Economies) (in Japanese): 203–27.

Toyoda, H. (1998) *Covariance Structure Analysis: A Primer* (Tokyo: Asakura Publishing) (in Japanese).

Toyoda, H. (2003) *Covariance Structure Analysis: Q & A* (Tokyo: Asakura Publishing) (in Japanese).

Ueta, K. (1996) *Environmental Economics* (Tokyo: Iwanami Shoten), (in Japanese).

Ueyama, S. (1998) *Reinventing Japan: A Review of Government Performance* (Tokyo: NTT Publishing), (in Japanese).

UNDP (1998) *Capacity Assessment and Development in a Systems and Strategic Management Context* (UNDP/BDP/Management Development and Governance Division), http://magnet.undp.org/Docs/cap/CAPTECH3.htm (accessed at 2 September, 2005).

UNDP/GEF (2003) *Capacity Development Indicators.* UNDP/GEF Resource Kit (no. 4), http://www.undp.org/gef/undp-gef_monitoring_evaluation/sub_undp-gef_monitoring_evaluation_documents/CapDevIndicator%20Resource%20Kit_Nov03_Final.doc (accessed at 2 September, 2005).

UNEP/WHO (1996) *Air Quality Management and Assessment Capabilities in 20 Major Cities*, Division of Environment Information and Assessment, UNDP and Urban Environmental Health, WHO (London: MARC).

Yamauchi, Y., H. Suzuki and S. Shibukawa (2000) 'Policy Design and Knowledge Management', *GLOCOM Review*, 5 (5): 1–17, (in Japanese).

Yanai, H., Kazuo Shigemasu, S. Maekawa, and M. Ichikawa (1990) *Factor Analysis* (Tokyo: Asakura Publishing) (in Japanese).

4
Development Stages Analysis for Social Capacity Development

Shunji Matsuoka, Metin Senbil and Yoshi Takahashi

Social capacity assessment and development stages analysis

Social capacity for environmental management (SCEM) can be defined as the extent of effectiveness in maintaining high levels of environmental quality and in managing environmental problems through the interplay of three social actors; namely, governments, firms and citizens. The core of the SCEM programme consists of two successive and complementary procedures: capacity assessment and capacity development. Capacity assessment (CA) broadly aims at estimating the existing social capacity. Based on the information obtained during capacity assessment, capacity development (CD) seeks to define the best approach to increase social capacity for environmental management. CA is a process of multi-stage analysis covering indicator development, actor–factor analysis, and development stages analysis. CA starts with a detailed account of the actors and their respective capacities in an actor–factor matrix design (see Chapter 3). To calculate the social capacity of a social actor, one needs pertinent indicators of social capacity, which is subject to indicator development (see Chapter 2). Based on the available indicators, actors and factors are tabulated with their existing and expected social capacities. Thus, the gaps between existing and targeted social capacities are highlighted to devise possible ways of ensuring optimum capacity development.

The initial stages of capacity assessment involving actor–factor analysis and indicator development are followed by development stages analysis (DSA). This completes capacity assessment and connects capacity assessment with studies on capacity development. In this framework, DSA follows an evolutionary approach toward improving the quality of the environment with three consecutive stages termed system making, system working, and self-management. Implicit in DSA is the understanding that environmental quality shows the first signs of improvement during the first two stages when social capacity is established and made effective. This is followed by the self-management stage when environmental quality is sustained (perpetually) at manageable levels. The stages differ from each other according to the factors (i.e., policies, knowledge and organization) associated with the social actors.

As regards the development stages, the first stage that involves system making is characterized by the steps necessary for healthy development. Thus, this stage is largely associated with institutionalizing environmental protection. The next stage, the system working stage, builds on the previous stage and entails the actual functioning of the institutions that were set up in the previous stage. The last stage is dependent on when society is believed to have reached a level where the social actors are fully functional and armed with checks and balances. In other words, at this stage, we believe that the social actors have become fully active and have acquired the potential to be effective on each other.

There are at least two ways of conducting development stages analysis while maintaining a connection with the analyses of the CA procedure that precedes it. In a fully connected system (integrated CA and CD approach; see discussions below), the analyses are conducted on the same data set. Thus, the earlier stages of CA are expected to provide inputs for DSA on areas such as current and target capacities as well as the effects of various policies on social capacity, thereby rendering DSA as an optimization problem. According to this approach, the best environmental programme is the one that minimizes the time as well as entailing costs required for transiting from one stage to another, especially from the system making to the system working stage. On the other hand, in a loosely connected system, constituent analyses may not be characterized by the input–output relationship as their subject (as is the case in this book; one might notice this by reviewing previous chapters). Thus, each part might concentrate on one single issue, leaving DSA as a fact finding problem that calls for statistical data analysis. The study conducted in this chapter uses a structural equations modelling technique to assess the relationships between social actors at different stages, and it takes into account a separate database from previous CA analyses to meet this purpose. A specially designed structure of relationships connects the three successive stages to each other by assuming the government as the key social actor in maintaining the institutions and realizing the social capacity for environmental management.

In the next section, we discuss the theory of development stages analysis as originally developed by Matsuoka *et al.* (2004). We set the stage by describing the social actors briefly along with their basic functions in the SCEM framework; thereafter, we give a detailed account of the stages of development of the social capacity for environmental management by devoting a separate subsection for each stage. In the same section, we also discuss the integration of the different development stages, the division between them, as well as the benchmarks needed for successfully completing any one of them and proceeding to another. In the same section, we also present country-based case studies that summarize the progress made by these countries in terms of the development stages. The reader is expected to note the difference between countries with different paths of social capacity development. We then discuss modelling the development stages analysis. Two modelling approaches are discussed, building on the previous discussion. We then focus on one of the two approaches discussed; namely, statistical data analysis. Using a structural equations modelling technique, we prepare a dynamic process model for development stages analysis. Estimation is carried out first in the system making stage, then in both the system making and system working stages, and finally in all the stages put together. Following this, we present the results of our estimation and analyze their implications for policy development.

Finally, we present a discussion on two interrelated themes: the path dependency of development stages and its implications for policy designs. Based on the theory established and inferences derived in previous sections, we posit that environmental policy-making within the SCEM framework is an evolutionary process with a high degree of path dependency. In other words, the initial state of social capacity is highly affective on the overall development path that unfolds eventually. One can thus view initial social capacity as the backdrop to the social capacity development scheme at the beginning of the system making stage. Thus, the success of any capacity development programme for improvements in the environment is highly dependent on the state of the social capacity at the time of the implementation of any policy associated with the different development stages of SCEM.

Theory of development stages analysis and related case studies

Matsuoka *et al.* (2004) proposed a development process for SCEM involving three consecutive constituent stages: system making, system working and self-management. Each development stage serves as yet another instance of interactions among the three social actors (i.e., government, firms and citizens) and the factors that characterize them at each stage. We first give a brief introduction to the social actors. Thereafter, we enumerate the development stages devised under SCEM. Each subsection covering the different development stages details one of the stages. The part dealing with theory ends with a discussion on the integration of the different stages. We also present case studies on development stage analysis conducted in four countries – China, Thailand, Indonesia and Mexico – to delineate the different stages of development in these countries.

The social actors

The government

The government has both the authority and the power to make and enforce laws. It also has the means to administer or supervise a specified jurisdiction. Modern political theory defines the government as a collective of three main powers: legislative (that makes laws), executive (that implements laws) and judiciary (that judges compliance with laws and dictates punishment when laws are broken). In a modern democracy, it is essential to put in place a system of separation of powers into three distinct branches of government that operate independently while acting as checks and balances upon each other. This separation is intended to prevent any one individual or small group from acquiring too much power for themselves and becoming despotic. In SCEM, the basic function of the government is reduced to establishing and maintaining institutions and organizations for environmental management. The government carries out these functions through its administrative, legislative and judicial bodies.

Firms

A firm is viewed as a collection of individuals or an organization that produces goods and/or services for the purposes of economic gain. The rationale for the existence of a firm and the scale of its operations is closely related to the existence of transaction and other related costs associated with contracted parties of the firm (Coase, 1937). It is assumed that firms pollute significantly in the system making stage as a result of insufficient institutions and lax and/or ineffective enforcement. However, the pollution caused by the activities of the firms begins to decrease after reaching its maximum during the system employment stage, and stays at manageable levels during the self-management stage. This is generally in compliance with the Environmental Kuznets Curve (EKC), which, however, is found to be applicable for only certain types of pollutants (e.g., SO_2 (Cole, 1999)).

Citizens

Citizenship is defined as a membership in a political community (e.g., a city or a state) and carries with it the rights to political participation; a person having such a membership in a society is a citizen. It is largely coterminous with nationality, which most often derives from place of birth (i.e., *jus soli*), and, in some cases, ethnicity (i.e., *jus sanguinis*). Citizenship often also implies working towards the development of one's community through participation, volunteer work and efforts to improve life for all fellow citizens. Citizenship, as explained above, is a collection of the political rights of an individual within a society. Citizenship derives from a legal relationship with the state. In SCEM, citizens are simply viewed as individuals or groups of individuals trying to achieve the most desired environmental conditions. Citizens or citizen groups can influence environmental issues insofar as they are aware of the environmental problems surrounding them. Citizens can stay abreast of issues concerning the environment by using the several information sources at their disposal. To collect information by themselves, citizens need secured rights to acquire information (e.g., existence of an information act) but, more importantly than even the very existence of such institutions, procuring and using (environmental) information requires different elements of human capacity generally gained by ways of education and/or affluence.

Development stages

Stage 1: system making

The first stage, the system making stage, is initiated by institutionalizing the environmental management system. Thus, the basic tenet of the system making stage is to establish the institutional and administrative framework needed for environmental management. Institutions are the collection of both informal (norms, customs, etc.) and formal (basic laws, statutes, property rights, contracts, etc.) rules governing actions and interactions (North, 1990). The history of institutional development suggests that it is path dependent and evolutionary rather than revolutionary (North, 1990). For institutions to be effective, a community needs strong administration as well as efficient judicial facilities to insure the proper functioning of the institutions.

Historical evidence suggests that the institutional and administrative frameworks established to address environmental issues might not be comprehensive initially; they might rather be sporadic and immature (see below; also, see Honda (2003) for reports on the developments in various developing Asian countries). Within the SCEM framework, it is expected that the effectiveness and efficiency of the institutions are expected to improve at the later stages owing to increased cooperation from both citizens as well as firms.

In the system making stage, in most cases, institutions are established in order to address general public concerns about various environmental issues instead of forming comprehensive institutions for an integrated environmental management system. Historically, public awareness on environmental issues has been a function not of a comprehensive outlook toward environmental problems, but rather of specific environmental problems and the resultant health problems affecting human or other living populations. The key element in this is the spread of pertinent information among the general public, triggering demands for proper action to prevent pollution and compensate for losses.

The well-known book *Silent Spring* by Carson (1965) testifies to this: the book adequately documents the health problems caused by exposure to agricultural chemicals such as DDT, and it ingeniously makes use of stories to reach out to the lay people. The public outcry that followed the publication and popularization of the book led to the phasing out of DDT from agricultural practices in USA as well as in other countries. The Love Canal (New York, USA) incident constitutes another typical example that sparked public protests and demands for proper actions to compensate for the environmental and health damages inflicted on human populations (Levine, 1982; Regenstein, 1982). In Japan, the outbreak of the Minamata disease in humans as a result of mercury poisoning via seafood caused a similar public reaction. These examples thus constitute a chain of sporadic events that keep the relevance of environmental problems at levels that cannot be ignored by elected officials aspiring to maintain their positions as representatives of the society.

Stage 2: system working

During the system working stage, the environmental management system seeks a new equilibrium induced by the institutional changes that emerge in the system making stage. Both citizens and firms become more active in this stage as a result of the institutional and organizational developments in the previous stage. Thus, the relations between the social actors are intensified. This stage is characterized by comprehensive environmental policy-making and the involvement of stakeholders other than the government (i.e., firms and citizens) in the decision-making processes. Thus, in this stage, the occasional legislative steps taken during the system making stage are combined to create generic laws (e.g., an environmental pollution act). Furthermore, different agencies related to different environmental issues are combined to constitute a ministry which enhances the scope of environmental protection and liaison among different units.

At the initial stages of economic development, the increasing scale of economic activities, as well as the changing composition of production from agricultural to industrial production, results in increased pollution. However, as incomes rise, the demand for high levels of environmental quality increases and more stringent environmental regulation leads to the replacement of old technologies of industrial production with less harmful ones. After reaching its peak, the level of pollution created by industrial production (Bai and Imura, 2000) decreases, thus justifying the trajectory suggested by EKC. The main hypothesis associated with EKC is that the emission of a certain class of pollutants (such as SO_2, related primarily to industrial production) first increases with production owing to scale effects. However, with the passage of time, emissions decrease as the per capita income crosses a certain threshold. This transformation may be attributed to a shift in the composition of production (from manufacturing to services) to a change in the techniques used to eliminate pollutants or to a technological change in the production processes (see Merlevede *et al.*, 2004). The latter two changes generally occur with the adoption of new policies. With economic development, economies also change structurally and move from industrial production to postindustrial economic activities. This puts a downward pressure on environmental problems caused by industrial production. Eventually, as the per capita income crosses a particular threshold level, income and composition effects outweigh the scale effect and the income–environment relationship tends to exhibit a downward slope.

Higher income levels are accompanied by an increased demand for a better environment in a system where environmental quality is assumed to be a normal good. This then points to the imperative for a new environmental outlook on the government's part. However, at this stage, the demand for environmental quality cannot be increased enough unless the public is well supplied with the pertinent information. This may potentially lead to a proliferation of organizations providing such information (e.g., non-governmental agencies or groups of scientists – some global examples are Greenpeace, the World Wildlife Fund, and the Union of Concerned Scientists).

Stage 3: self-management

Finally, during the self-management stage, the SCEM is supposed to reach a new partial equilibrium that requires cyclic interactions between actors. In this stage, society is supposed to define the environmental problems that it faces and devise relevant solutions by using the synergy between social actors and the well-established information channels existing in the new institutional environment.

For example, in Japan during the early 1980s, when the minister in charge of environmental affairs declared his formal opposition to the antipollution measures instituted by the very active industrial lobby in central government, he faced strong opposition from the public and local governments as well as international trading partners, and had to change his position and reconfirm his commitment to environmental protection (Lovei and Weiss, 1998). This case is a typical example of policing to prevent deviations from societal commitments to environmental protection. The actors in this stage exert mutual influence on the functioning of each other. While democratic institutions allow citizens and firms to lobby for their priorities at the governmental level, firms try to find ways to attract consumers (i.e., citizens) by advertising voluntary programmes in those areas of environmental management that draw their involvement at the organizational level. The ISO 14000 series, British Standards 5750 and 7750, and EMAS[1] are examples of voluntary environmental management schemas that can be adopted by firms. Such compliance is advertised well by means of eco-labelling in order to attract customers who are willing to pay higher prices for environmentally friendly products.[2] This pattern can also be discerned in the relationship between firms and the government. The government can offer incentives to induce firms to adopt environmentally friendly ways of operating.

Integration of development stages

Each development stage, in effect, represents a non-overlapping snapshot of social capacity development, which is a continuous phenomenon. In this discrete version of social capacity development, social actors are assumed to improve in terms of social capacity at each development stage. The general characteristics of the development stages within the SCEM framework are distilled from the above discussions and presented in Table 4.1. According to Table 4.1, institutions are established in the

Table 4.1 Characteristics of SCEM development stages

Stages	System making	System working	Self-management
Definition	The stage when institutions are established	The stage when institutions start working effectively	The stage when intensive cooperation between actors starts
Environmental issues	Poverty-related environmental issues, industrial pollution	Industrial pollution	Consumption-related issues
Industrial pollution	Increasing	Turning point	Decreasing
Benchmarks (essential)	1 Environmental law 2 Ad hoc administration for environment 3 Collection and distribution of environmental information	1 Regulation	1 Comprehensive environmental approach
Benchmarks (important)	1 Negotiation 2 Mass media	1 Negotiation, adjustment and cooperation	1 Voluntary participation

Source: Adapted from Matsuoka (2003).

system making stage and their practices are made effective in the system working stage. When any environmental law is enacted, the related administrative systems are generally established after a while because of constraints related to time, budget and skilled personnel. The time required for the administrative systems to start functioning at full capacity constitutes the core of social capacity development during the system working stage.

However, the progress from the system making to the self-management stage does not necessarily alleviate environmental degradation in general and environmental pollution in particular, as during this process society also changes in character. For instance, Bai and Imura (2000) have proposed three types of pollution: poverty related issues (Type I), rapid growth issues (Type II) and consumption and lifestyle related issues (Type III). Type I pollution is expected to decrease smoothly with economic development, Type II pollution is expected to follow the path charted out in EKC and Type III pollution is only expected to increase. This classification fits the list of environmental issues enumerated in Table 4.1.

At least two types of benchmarks are available to assess whether a society has passed a certain development stage. The first is classified as the essential benchmark that must be met in order to fulfil the basic requirements of a development stage; the second is classified as the important benchmark that enhances the efficacy of a given development stage, increasing the pace and strengthening the impact of the essential benchmarks in the later stages as a result of high path dependency in development stages.

Country-based case studies

Based on the basic concept as discussed above, the development stage of SCEM in the five selected countries (including Japan as a reference case) is summarized in Figure 4.1.

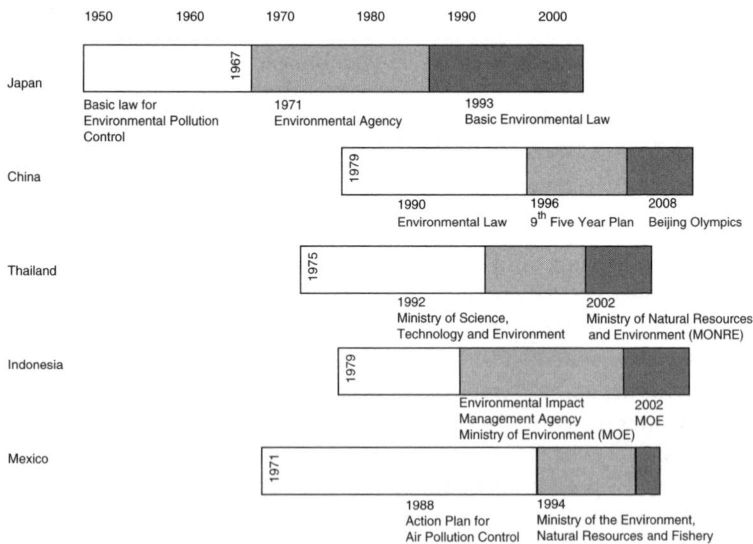

Figure 4.1   SCEM development stages of selected countries
Source: Evaluation Team on Environmental Cooperation, JASID (2003).

China

In China, both environmental law as well as administrative measures were satisfactorily established in the 1990s. The *China Environment Yearbook*, which is the official report on the state of the environment in China, has been in publication since 1990, having upgraded its standing since 1994. This is evidence enough of the fact that the system making stage in China was over in the mid-1990s and that the first half of the 1990s thus covered the final phase of that stage. With the amendments to the Air Pollution Control Act in 1995 and the beginning of the ninth Five-year Plan period in 1996, China implemented effective countermeasures to combat pollution and entered the system working stage in the second half of the 1990s. Since industrial SO_2 emissions in China reached their peak in 1996, there is a possibility that China reached the turning point in terms of pollution control in the latter half of the 1990s. The process of developing social capacity appears to have taken extensive roots in China; the government, firms and citizens, acting as a single body, appear to be actively promoting environmental management in the lead up to the Beijing Olympic Games to be held in 2008 and the Shanghai International Exposition to be held in 2010, and the country seems to have started its transition from the system working to the self-management stage.

Thailand

Thailand has well organized environmental laws, administrative apparatus, and channels of information, and the country shifted to the system working stage from the system making stage in the mid-1990s. However, it has taken a considerable while to set up the system working stage in the social environmental management system (SEMS) because of the social and economic unrest caused by the currency crisis in 1997. Furthermore, in Thailand, the early phase of the system working stage coincided with the period of governmental reorganization involving the restructuring of the former Ministry of Science, Technology and Environment (MOSTE) into the present Ministry of Natural Resources and Environment (MONRE) following the promulgation of the new Constitution in 1997, the enforcement of the Decentralization Plan and Process Act in 1999 and the restructuring of the ministries in October 2002.

Indonesia

Environmental law and administration in Indonesia were developed in the late 1980s and early 1990s. Nevertheless, Indonesia is lagging behind in terms of the distribution of environmental information. There is no nationwide monitoring network; neither is there a system of periodical dissemination of information related to the state of the environment. Under these conditions, the country appears to have been at a standstill in the final phase of the system making stage since the beginning of the 1990s. Furthermore, Indonesia went through social and economic turmoil owing to the change of the Suharto administration along with the currency crisis in 1997, the independence movement of East Timor and the restructuring of all administrative bodies. This restructuring began with the establishment of the new Ministry of the Environment and BAPEDAL (Environmental Impact Management Agency) (January 2002) – which took over from the State Ministry of the Environment in the wake of revisions in central ministries – and the enactment of the Decentralization Act (2001). Under unstable administrative conditions such as these, Indonesia may yet stagnate in the final phase of the system making stage.

Mexico

In Mexico, environmental laws and administration were developed from the end of the 1980s to the middle of the 1990s (SEMARNAP, Ministry of the Environment, Natural Resources and Fishery, established in 1994). The distribution of environmental information among the general public was also introduced at around the same time. The development of the social environment management system in Mexico was completed in the mid-1990s and the country now appears to be shifting to the self-management stage from the system working stage. However, Mexico City witnessed a turning point in SO_2 emission between 1992 and 1993, and, according to this data, the system working stage had already started in the first half of the 1990s. Furthermore, the Action Plan for Air Pollution Control (in 1988) and the Integral Program for Air Pollution Control (PICCA, 1990–95) had also been implemented around the same time. It can thus be asserted that the country had already been in the system working stage as well as the final phase of the system making stage simultaneously in the late 1980s. This analysis suggests that the foundation of CENICA (National Center for Environmental Research and Training of Mexico) in 1992 was a little too late to contribute significantly to Mexico's social capacity development for environmental management.

Modelling the dynamic stages analysis

Generally, DSA as theorized in the previous section can be modelled in two ways that tally with two different views of the CA procedure within the SCEM framework. The first of these views CA as a fully integrated system characterized by input–output relationships for both the constituent analyses as well as the capacity development procedure (Figure 4.2) – hence the name integrated capacity assessment and capacity development approach.

In this integrated system, the entire process of CA is divided into two groups of analyses. The first group includes the basic actor–factor and indicator development analyses. The output of the first group of analyses is the value of social capacity, which becomes the primary input for the DSA. Interacting with environmental programme development, DSA refines the target social capacity that was previously determined in the total system analysis (see Chapter 1 for detailed information on total system analysis). This system, which takes into account the current value of social capacity and the costs of the different programmes, posits DSA as no more than an optimization problem.

The second view of DSA projects it as a statistical estimation problem. This is mainly caused by the loose integration of the constituent stages that might emerge when different forms of analyses are independently implemented. Based on the trends observed in a number of countries, statistical estimation is carried out by assessing the theoretical relationships between different social actors using the factors employed by them at different stages. In this process, a social actor is viewed as a latent variable at a certain SCEM stage, and we assume that this latent variable is observed by the factors associated with it.

We adopt the second approach. We apply the structural equation modelling (SEM) technique to estimate the structural parameters of the hypothesized effects among the social actors, pursuing the line of modelling proposed by Senbil *et al.* (2005) but using a different estimation strategy.

Dynamic process model of the development stages analysis

An essential hypothesis for the SCEM procedure proposed by Matsuoka *et al.* (2004) is that it is temporally sequenced in stages (i.e., system making, system working and self-management. However, transition from one stage to another in the original form of SCEM lacks explicit detailing. Following the analysis of Senbil *et al.* (2005), the current model addresses the transition from one stage to another by linking the three stages with the government. We regard the government as the key actor since it has the power to establish formal institutions and to maintain them subsequently. In keeping with this, Figure 4.3 shows the proposed model structure with all the stages linked by the government at different junctures. We assume that the government affects itself in the early stages

110

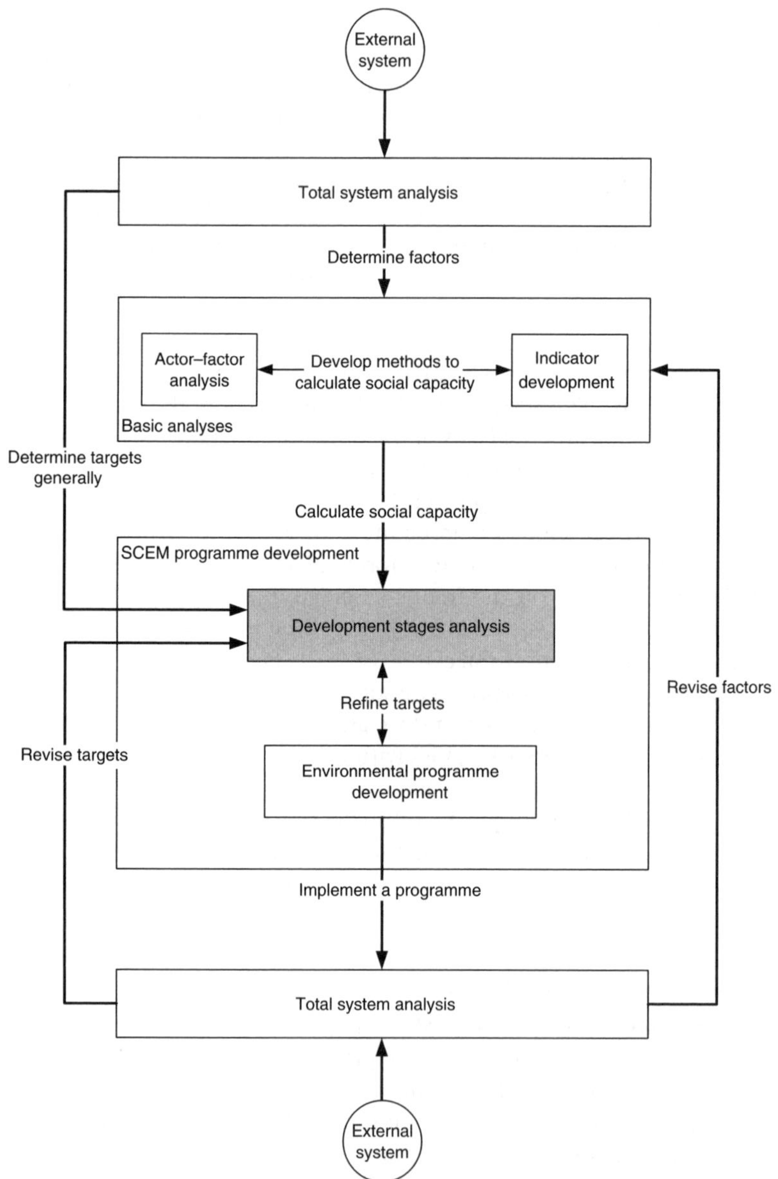

Figure 4.2 Social capacity assessment and capacity development approach

(i.e., G1→G2) with the introduction of new legislation. The system working stage is linked to the self-management stage, with the government exercising its influence on the other actors.

In this model, as mentioned above, the onset of the system making stage coincides with the formation of relevant formal institutions, which become effective during the system making stage. Information about international treaties is used to shape domestic legislation for the purpose of institution building. The economy is assumed to develop and the per capita income expected to grow in the system working stage. Thus, as Figure 4.3 demonstrates, there is an improvement in the quality of life with the stages progressing from left to right. Citizens begin to exert pressure on the private sector (firms) to comply with the standards needed to maintain or improve their environmental performance in order at least to ensure stability in the quality of life. The private sector, in turn, has to cope with this pressure and respond suitably in order to maintain its material well-being. Besides, they are subjected to more pressure from the government owing to its commitment to international treaties seeking to improve environmental quality.

During self-management, the actors start influencing one another actively and the system works in terms of two-way relationships. To sum up, this model uses a multi-stage development framework for SCEM. By using the government as a catalyst for transiting from one stage to another, the model modifies the basic structure available for SCEM to extend the basic fragmented structure to a dynamic process.

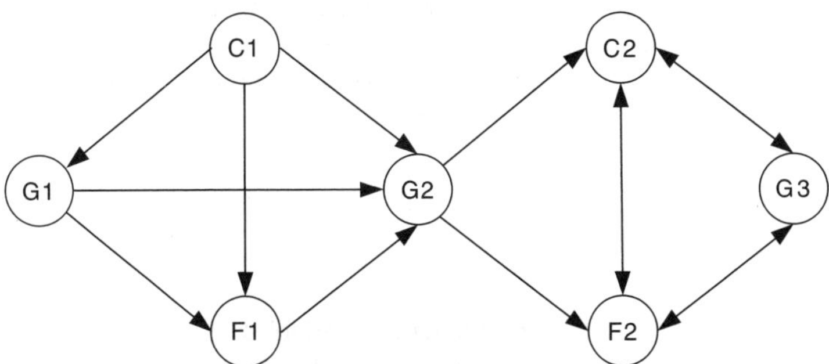

Figure 4.3 Dynamic process model of social capacity for environmental management

Source: Adapted from Senbil *et al.* (2005).

Estimation strategy

Estimation of the model is carried out in stages by reducing the original data size in each stage (Table 4.2). The three estimation stages overlap with the three development stages. During the first stage of estimation, we take into account the full data size covering all the countries in the database. This confirms the assumption that all the countries are either in or have completed the system making stage. With the parameters of the system making stage set to their estimated values, the second estimation stage appends the system working relationships to the system making relationships. Thus, the model includes all the relationships from the system making to the system working stages. However, the parameters to be estimated at the second stage are those of the system working relationships only. The enlarged model is estimated with a reduced data size covering all countries except poor countries (according to the World Bank classification) in order to exclude the countries that have not yet emerged from the system making stage from the process of estimation. This helps us in thwarting the bias that might be induced by their inclusion. The procedure of using the estimated parameters as fixed during the previous stages is repeated in Step 3, as shown in Table 4.2. During the last stage – self-management stage, we estimate the reciprocal effects between social actors. In Step 3, the data is further reduced to cover only the rich countries.

As mentioned above, the model proposed in this study is formulated as an SEM with latent variables. Thus, it is in order to provide a brief introduction to the latent variables used in SEM. SEM with latent variables consists of two constituent models:[3] the structural model and the measurement model (see Bollen, 1989). The structural model captures the relationships between the latent variables, and the measurement model relates latent variables to observed variables.

Consistent with the LISREL model (see Jöreskog and Sörbom, 1993), let the structural equations be expressed as:

$$\eta = \mathbf{B}\eta + \mathbf{\Gamma}\xi + \zeta \tag{4.1}$$

where
η is an $m \times 1$ vector of latent endogenous variables,
ξ is an $n \times 1$ vector of latent exogenous variables,
$\mathbf{B}$ is a $m \times m$ matrix of coefficients,
$\mathbf{\Gamma}$ is a $m \times n$ matrix of coefficients, and
ζ is an $m \times 1$ vector of random errors.

Table 4.2   Estimation technique of dynamic process model

Step 1   • Estimate
parameters: $[\alpha_1]$,
$[\alpha_2]$ and $[\beta_1]$

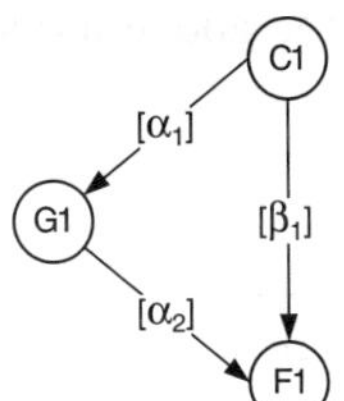

Step 2   • Fix the values of
α_1, α_2 and β_1 as
estimated in
Step 1, enlarge
model to include
system working
variables

• Estimate
parameters $[\alpha_3]$,
$[\alpha_4]$, $[\alpha_5]$, $[\beta_2]$, $[\gamma_1]$

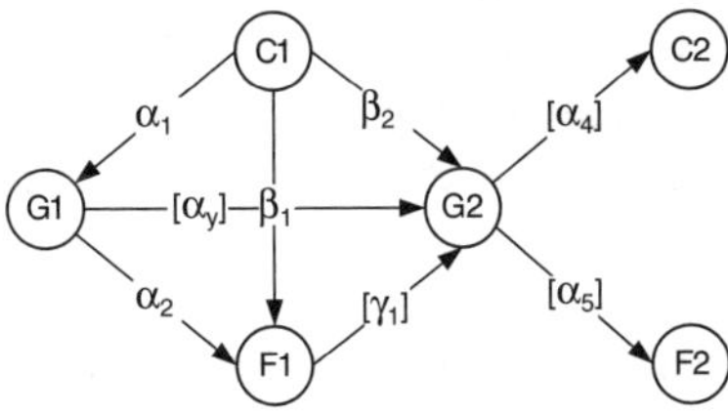

Step 3   • Fix the values of
α_1, α_2, β_1, α_3, α_4,
α_5, β_2, γ_1 as
estimated in Step
2, enlarge model to
include self
management
variables

• Estimate
parameters $[\gamma_2, \beta_3]$,
$[\beta_3, \alpha_7]$, $[\gamma_3, \alpha_8]$

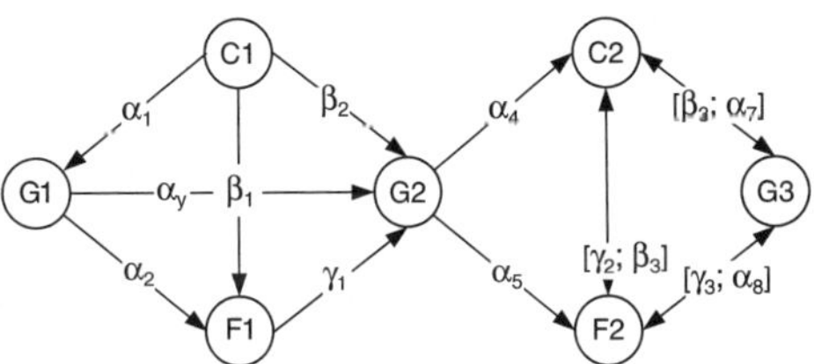

The measurement equations can be expressed as follows:

$$y = \Lambda_y \eta + \varepsilon \tag{4.2}$$

$$x = \Lambda_x \xi + \delta \tag{4.3}$$

where
y is a p×1 vector of observed endogenous variables,
x is a q×1 vector of observed exogenous variables,
Λ_y is a p×m matrix of coefficients,
Λ_x is a q×n matrix of coefficients,

ε is a p×1 vector of random errors, and
δ is a q×1 vector of random errors.

The stochastic errors of the above model are assumed to satisfy the following conditions:

1   ε is uncorrelated with η,
2   δ is uncorrelated with ξ,
3   ζ is uncorrelated with ξ, and
4   ε, δ, and ζ are mutually uncorrelated.

The covariance matrices of ξ, ζ, ε and δ are denoted by $\Phi(n \times n)$, $\Psi(m \times m)$, $\Theta_\varepsilon(p \times p)$ and $\Theta_\delta(q \times q)$, respectively.

The above system of equations is estimated by equating the observed covariance matrix of $\mathbf{y}$ and $\mathbf{x}$ with the covariance matrix of these variables implied by the model (see Bollen, 1989). Thus

$$S = \Sigma(\Theta) \tag{4.4}$$

The covariance matrix on the left-hand side is the sample covariance matrix; the one on the right hand side is the model-derived covariance matrix obtained by applying the expectation operator on the equation system above. The model-derived covariance matrix using $\mathbf{y}$ and $\mathbf{x}$ is as follows:

$$\Sigma = \begin{pmatrix} Var(\mathbf{y}) & Cov(\mathbf{y},\mathbf{x}) \\ Cov(\mathbf{x},\mathbf{y}) & Var(\mathbf{x}) \end{pmatrix}$$
$$= \begin{pmatrix} \Lambda_y A \left(\Gamma\Phi\Gamma' + \Psi\right) A'\Lambda'_y + \Theta_\varepsilon & \Lambda_y A\Gamma\Phi\Lambda'_x \\ \Lambda_x\Phi\Gamma'A'\Lambda'_y & \Lambda_x\Phi\Lambda'_x + \Theta_\delta \end{pmatrix} \tag{4.5}$$

where $A = (I - B)^{-1}$.

The maximum likelihood method is applied to equate the terms in Equation 4.5 with the assumption that y and x follow a multivariate normal distribution and that consequently the variance-covariance follows the Wishart Distribution. By replacing the sample covariance S with the population covariance Σ in equation 4.4, we find that the maximization of the log-likelihood function (see Bollen, 1989: 131–5) is equivalent to the minimization of the following function:

$$F_{ML} = \log|\Sigma(\Theta)| + tr\left(S\Sigma^{-1}(\Theta)\right) - \log|S| + (p+q) \tag{4.6}$$

where S is the sample covariance matrix.

Data and variables

The data used in this study are collected from three sources, all of which are available on the Internet. Data about environmental treaties were obtained from the 'Environmental Treaties and Resource Indicators' web page[4] maintained by Columbia University. Data about all other variables were obtained from the Environmental Sustainability Index database,[5] also maintained by Columbia University, and the UN GeoData portal[6] maintained by the UN Environment Programme. In fact, these data sources themselves supply information collected from several different sources. We have also provided the original sources of these data below. For more information, the reader is referred to the ESI portal on the Internet.

The observed variables used in the model are selected in compliance with the theoretical foundations of SCEM. As the activities of the government are initially associated with institution building, the observed variable that is available to us, vis-à-vis the government, in the system making stage (G1) is the set of international treaties concerning environmental protection that has been the foundation for related domestic laws. Between 1868 and 1999, a total of 464 pertinent international treaties were signed. The latent variable G1 is measured in terms of the number of treaties that are ratified by different countries. We weighted this original variable with the state of the rule of law in each country in order to better represent the efficacy of the environmental institutions. The rule of law variable is originally supplied by the World Bank, and is measured in terms of the perceptions of crime, the effectiveness of the judiciary, and the enforceability of contracts. The original variable for the rule of law is rescaled to take values between 0 (lowest) and 1 (highest).

The Agenda 21 initiatives launched per million people at the system working stage and knowledge creation exercises covering environmental science, technology and policy at the self-management stage are considered to be variables that closely represent the behaviour of the government at the later stages of this model (G2 and G3). Agenda 21 initiatives generally evaluate the cooperative projects launched by local governments and civil society organizations. This variable is obtained by dividing the number of Agenda 21 initiatives undertaken by the population of a country. The original source for this variable is the International Council for Local Environmental Initiatives.[7] In the last stage of this model, we use the variable related to knowledge creation in environmental science, technology and policy, which is one of the most complicated variables used in this study. It is developed as a part of the Knowledge Divide Project undertaken by the Yale Center for Environmental Law and Policy[8] using data from publications that appeared in nine highly ranked, peer reviewed journals on environment, ecology and related subjects. Three regressions are conducted with dependent variables based on the number of publications per author per million population, publications about foreign countries and publications per unit of area; the residuals of these regressions are ranked and aggregated to compile the variable.

In the system making stage, citizens (C1) are associated with evenness of income distribution (i.e., evenness of income distribution = $100 - \text{GINI}$) and the Human Development Index (HDI). HDI is calculated by the United Nations and it is a measure of the various dimensions of social welfare: a long and healthy life (life expectancy at birth), knowledge (2/3 adult literacy rate, 1/3 gross enrolment rate) and a reasonable standard of living (PPP corrected GDP per capita). The first and the last dimensions of HDI are converted to take values between 0 and 1; hence, the value of HDI ranges between 0 and 1. When the value of HDI tends toward 1, it represents a positive change in the quality of life of the citizens of a country.

In the last stage, the latent variable for citizens (C2) takes values from observed variables related to education, including the primary education enrolment rate among females and the gross tertiary education enrolment rate. Both variables are primarily formulated by UNESCO. The first variable gives the percentage of females within the same age group who have completed primary education, thus supplying information about the most affected population segment in terms of education and the general social and economic situation in the country. The second variable measures the percentage of population enrolled in tertiary education among the relevant population, thus completing the whole picture in terms of the continuation of education among the citizens in a country.

Firms in the system making stage (F1) are assumed to be the source of pollution. Therefore, we use two observed variables related to anthropogenic SO_2 emissions and coal consumption per unit of populated land area (km^2). The first variable is obtained from various international agencies.[9] The data for the most recent year for each country is extracted from these sources. The second variable on coal consumption is obtained from the US Energy Information Agency. The populated land areas are assumed to be inhabited by more than five people and their population is computed by using a gridded population dataset. Later, in the self-management stage, we note the number of ISO 14001 certificates awarded to firms per billion dollars of PPP corrected GDP. In this setting, the system making and self-management stages overlap, since firms and citizens are used as participants in both the stages. Interactions between the firms and citizens come into existence during self-management.

Table 4.3   Descriptive statistics for observed variables

	Observed variable		N	Min.	Max.	Mean	Std Dev.
1	International agreements accepted as domestic law		193	1	241	70.61	45.47
2	Rule of law (reduced to scale between 0 and 1)		143	0.00	1.00	0.44	0.26
3	Weighted international agreements accepted as domestic law (1 × 2)	G1	143	0.00	218.39	44.03	51.13
4	Average GINI 1960–2002 (subtracted from 100)	C1	120	29.34	75.56	59.67	10.57
5	Average of HDI 1975–2000	C1	162	0.26	0.91	0.66	0.18
6	Anthropogenic SO_2 emissions per unit of populated land area	F1	139	0.00	21.39	1.90	3.24
7	Coal consumption per unit of populated land area	F1	141	0.00	18.94	1.41	3.22
8	Local Agenda 21 initiatives per million people	G2	104	0.01	130.28	5.16	15.96
9	Number of ISO 14001 certified companies per billion dollars GDP (PPP)	F2	140	0.00	41.51	1.07	3.77
10	Gross tertiary enrolment rate	C2	120	0.57	77.62	26.78	21.69
11	Primary education completion rate among females	C2	129	39.00	100.00	90.77	15.02
12	Knowledge creation in environmental science, technology and policy	G3	73	1.67	74.67	39.18	17.06

Table 4.3 presents the descriptive statistics for the above-mentioned variables.

Estimation results and implications

Figure 4.4 presents the estimated structural relationships (standardized) derived by the maximum likelihood method.[10] According to the estimation results, there is a predominance of positive relationships between social actors at the later stages, reflecting a degree of path dependence in the overall progress of the development stages. Note that the capacity of the firms is assessed with respect to their inability to emit fewer pollutants

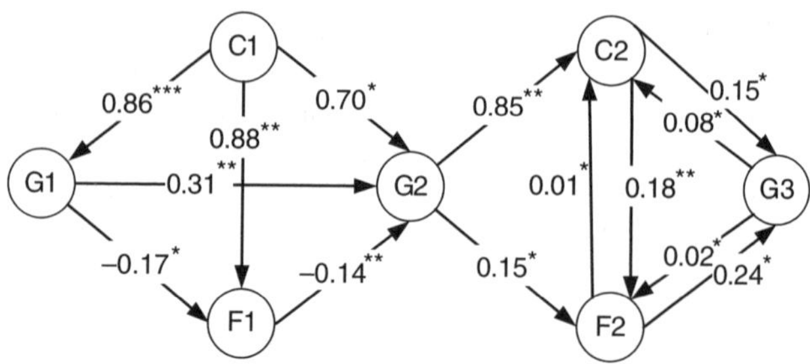

Figure 4.4   Estimated structural relationships

Note: Level of significance is *** if p ≤ 0.01,** if p ≤ 0.05, and * if p ≤ 0.10.

and their technological inability to consume fewer environmentally polluting inputs. Thus, the government has a significant effect in the reduction of the inability of firms to generate less pollution. In the later stages, the social actors influence each other's capacities, thereby increasing the overall social capacity.

The above results present several implications for improvements in social capacity. The first noteworthy observation is that the consideration of the well-being of the general population (as represented by C1) exerts a strong influence on central government and induces the government to be proactive in environmental issues, but it falls short of influencing the private sector in terms of environmental compliance. Civil society (as represented by C2) exerts a greater influence on the government as well as the private sector than the reciprocal effects of the latter two entities on it.

However, it should be noted that the model presented here deals with only one snapshot of the society and can be an over-simplification of the capacity of social actors as well as society as a whole. Thus, different results are possible with different observed variables that would be considered to be better indicative of the social capacity of a social actor at a certain development stage. For example, as regards the observed variables, some of them might account for other social actors as well as other development stages, beyond their immediate scope. For instance, the observed variable pertaining to knowledge creation in environmental science, technology and policy would also be considered as an observed variable for the citizens. But it is more appropriate to associate this variable with the government since it is the main body that supports research organizations and universities and has more authority in earmarking funds for research. Similarly, Agenda 21 initiatives could also be considered to be within the realm of the citizens. But without support from local governments, these initiatives cannot be realized, thus testifying to the degree of governmental authority in initiating projects and sustaining them within the Agenda 21 framework.

The model will also be more meaningful if the above system of analyzing development stages is integrated with the actual state of the environment in the various stages. However, this is difficult to implement since the model represents the logical separation of the development stages and not the flow of calendar time. It should also be noted that a statistical estimation of DSA comes with the basic limitations inherent in statistical estimation procedures in terms of model identification and the convergence of the loglikelihood to its best value.

Towards the creation of an effective environmental policy

One of the prominent features of DSA is the path dependent nature of social capacity. The capacity achieved in any given stage of development is strongly contingent upon the capacity achieved in the previous stages and the initial status of the capacity development activities. Thus, societies that follow certain routes of development end up with different results in terms of the final social capacity achieved. Furthermore, the time taken to achieve environmental goals may differ with respect to the initial status of social capacity.

It is an implicit assumption in DSA that once all the benchmarks for system creation are achieved, the transition from system making to system working is generally rapid. However, this happens more quickly in societies with high social capacity than societies with low initial levels of social capacity. When this phenomenon is considered along with the fact that environmental policy-making is a highly evolutionary process, we can derive certain implications for effective environmental policy-making.

For example, let us take the case depicted in Figure 4.5. The lines in Figure 4.5 represent the development of social capacity as the society proceeds from the system making stage to the self-management stage. The lines diverging after the system making stage represent business-as-usual (BAU) (the lower line) and capacity development programmes (the upper line). In the BAU case, society is expected to complete the essential benchmarks but not necessarily the important benchmarks. Thus, in this case, society takes longer to improve in terms of social capacity than in the case when any capacity development programme is initiated. In this setting, both lines show the boundaries of capacity development. The area bordered by the two lines and the two end points of the development

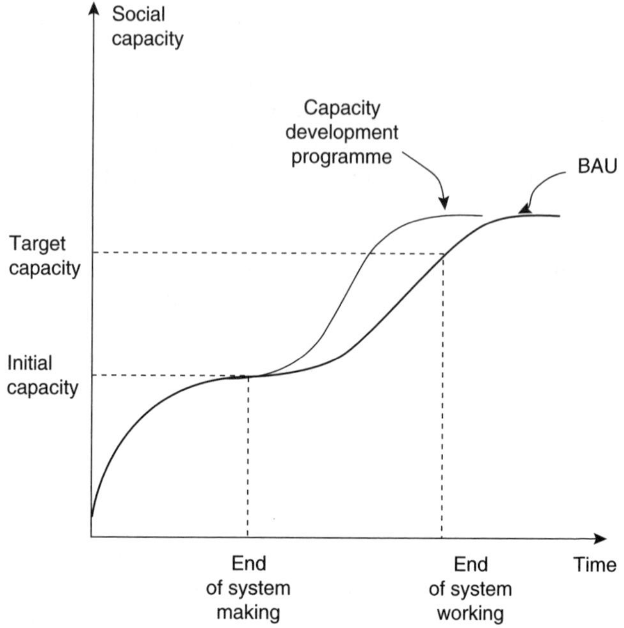

Figure 4.5 Social capacity development programme

stages – i.e., the end points of the system making and system working stages – represents the area in which environmental policies are developed and grouped together as environmental policy programmes. We name this area the environmental programme development area.

An environmental programme is a collection of different projects related to environmental management. The aim of an environmental programme is to enhance the effectiveness of the environmental policies instituted by society through interrelated social actors. Thus, the information that can be obtained by using DSA is highly important for the efficiency and effectiveness of environmental programmes. For instance, the information about the path and initial point dependencies can be instrumental in assessing the success of different programmes with regard to the savings made in time and money.

Notes

1 Eco-Management and Audit System.
2 For example, the European Union uses the flower symbol for eco-labelling.

3 Note that we only report the structural parameters in this study due to limitations of space.
4 http://sedac.ciesin.columbia.edu/entri/
5 http://sedac.ciesin.columbia.edu/es/esi/
6 http://geodata.grid.unep.ch/
7 http://www.iclei.org/
8 http://www.yale.edu/envirocenter/
9 United Nations Framework Convention on Climate Change (UNFCCC), Organization for Economic Cooperation and Development (OECD) and the Intergovernmental Panel on Climate Change (IPCC).
10 Goodness of fit statistics are as follows: in Step 1, $df = 3$, chi-square $= 2.54, p = 0.47, AGFI = 0.98$; in Step 2, $df = 34$, chi-square $= 169.01$, $p = 0.00, AGFI = 0.73$; in Step 3, $df = 48$, chi-square $= 197.46, p = 0.00$, AGFI $= 0.75$.

References

Bai, X. and H. Imura (2000) 'A Comparative Study of the Urban Environment in East Asia: Stage Model of Urban Environment Evolution', *International Review for Environmental Strategies*, 1 (1): 135–58.

Bollen, K. (1989) *Structural Equations with Latent Variables* (New York: John Wiley & Sons): 514.

Carson, R. (1965) *Silent Spring* (Harmondsworth: Penguin Books): 317.

Coase, R. (1937) 'The Nature of the Firm', *Economica*, 4: 386–405.

Cole, M. (1999) 'Limits to Growth, Sustainable Development and Environmental Kuznets Curves: An Examination of the Environmental Impact of Economic Development', *Sustainable Development*, 7: 87–97.

Evaluation Team on Environmental Cooperation, Japan Society for International Development (JASID) (2003) *Environmental Center Approach: Development and Social Capacity for Environmental Management in Developing Countries and Japan's Environmental Cooperation* (Tokyo: Japan International Cooperation Agency): 78.

Honda, N. (2003) 'Evaluating Social Capacity Development for Environmental Management in Developing Countries: Case Studies', *IDE Spot Survey: Social Capacity Development for Environmental Management in Asia*, Institute of Developing Economics (IDE-JETRO): 23–33.

Jöreskog, K. and D. Sörbom (1993) *LISREL 8: User's Reference Guide* (Chicago: Scientific Software International): 378.

Levine, A.G. (1982) *Love Canal: Science, Politics, and People* (Lexington, MA: Lexington Books): 263.

Lovei, M. and C.J. Weiss (1998) 'Environmental Management and Institutions in OECD Countries: Lessons from Experience', *World Bank Technical Papers*, 391.

Matsuoka, S. (2003) 'Social Capacity Development for Environmental Management', *IDE Spot Survey: Social Capacity Development for Environmental Management in Asia*, Institute of Developing Economics (IDE-JETRO): 7–21.

Matsuoka, S., S. Okada, K. Kido and N. Honda (2004) 'Development of Social Capacity for Environmental Management and Institutional Change', Proceedings of International Conference on Social Capacity Development for

Environmental Management and International Cooperation in Developing Countries, Hiroshima, Japan.

Merlevede, B., T. Verbeke and M. De Clercq (2004) 'The EKC for SO_2: Does Firm Size Matter? Ghent University Working Paper 2004/218.

North, D.C. (1990) *Institutions, Institutional Change and Economic Performance* (New York: Cambridge University Press): 152.

Regenstein, L. (1982) *America the Poisoned* (Washington, DC: Acropolis Books): 414.

Senbil, M., A. Fujiwara and J. Zhang (2005) 'Dynamic Process Model of Social Capacity Development for Environmental Management', *Journal of International Development and Cooperation*, Hiroshima University, 11: 31–50.

5
Designing Social Capacity Development Policy for Environmental Management

Shunji Matsuoka, Satoru Komatsu, Yumi Fukuhara,
Atsushi Ohno and Teppei Yamashita

Policy design based on social capacity assessment

Capacity assessment is an important methodology for aid effectiveness and capacity development. It contributes to helping developing countries to formulate strategies for sustainable development.

Social capacity assessment (SCA) is a catalyst, enabling developing countries to create their own development strategies. The SCA framework is combined with and integrated into planning capacity development policy, and this is followed by the assessment procedure discussed in Chapters 1–4. This chapter provides the conceptual framework of SCA based policy design, environmental policy design structure, their application to urban environmental management and an outlook for international environmental cooperation.

Analytical steps in policy design for social capacity development (SCD)

This part of the chapter represents the analytical steps for policy design structure. These steps comprise indicator development, actor–factor analysis and development stage analysis. It is also necessary to consider the scoping of environmental issues before implementing these steps.

Indicator development

Indicator development, discussed in Chapter 2, shows the current capacity of each Asian country. Social capacity for environmental management (SCEM) indicators enable us to determine the aggregate capacity of a country. It also enables the determination of the proportion of the country's capacity allocated to the government, firms and citizens. Indicator development clearly presents a country's capacity as well as the actors' capacity. In addition, this step enables us to make a country-by-country comparison of the SCEM situation.

Actor–factor analysis

Actor–factor analysis, discussed in Chapter 3, shows the detailed elements of capacity using three actors (government (G), firms (F) and citizens (C)) and three factors (policy and measure (P), organizational resources (R) and knowledge and technology (K)). The three actors focused on the stakeholders mentioned in Chapter 1. The present analysis is a detailed analysis of indicator development, which provides the country-specific indicators of SCEM and the actors' capacity.

Development stage analysis

Development stage analysis traces the development path of the capacities in each country. Assume that a country is now at the system making stage identified by development stage analysis. Due to path dependency – that is, due to the fact that the capacity development process is highly dependent on the country's institutional structure or historical background – the country takes a long time to reach the critical minimum. In this case, the critical minimum means the lowest level of capacity required to shift to the system working stage.

SCA-based policy design

The SCA based policy design structure is based on the steps discussed above. Indicator development numerically evaluates countries' and actors' capacities. It also enables a country to recognize its current social capacity and the actors' capacity. This analysis enables a country based comparison regarding the SCEM value. Actor–factor analysis enables one to make a detailed analysis of SCEM on the basis of the actors' and factors' capacity. This analysis reveals which factors or actors significantly affect the country's social capacity development. Figure 5.1 shows the current capacity of a country defined by indicator development and actor–factor analysis.

Development stage analysis traces the development path of each country from the initial point to the current point and predicts the future path of the country's capacity. This analysis also discusses the stages of SCEM, which are defined by a benchmark. Figure 5.2 shows the example of a country that is currently in the system working stage.

The SCA based policy is implemented for capacity development, which also enables a country to proceed to the next stages. Under this policy, aid encourages the country to achieve faster capacity development.

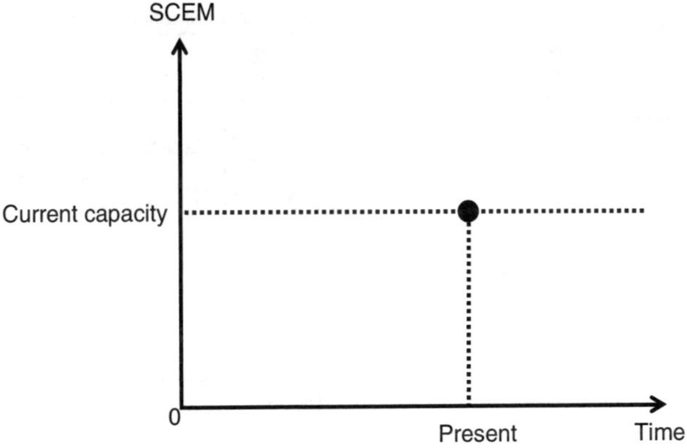

Figure 5.1   Indicator development and actor–factor analysis

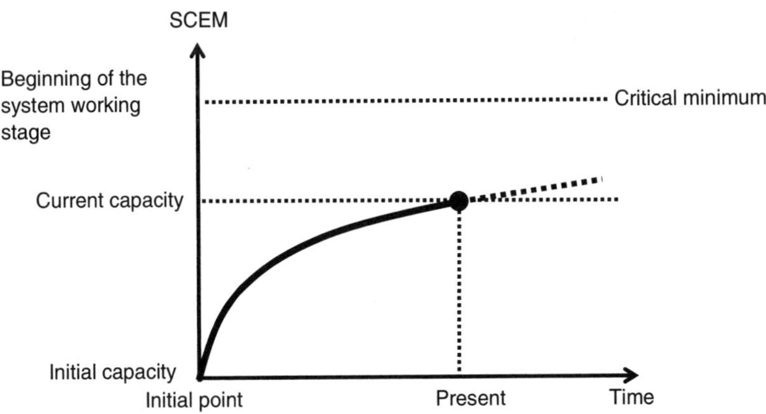

Figure 5.2   Development stage analysis

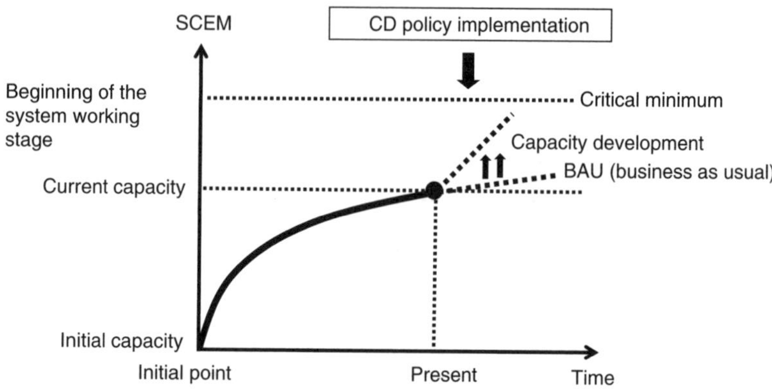

Figure 5.3   SCA-based policy design

Figure 5.3 shows that if the appropriate projects under SCA can be formulated, it can be expected that the capacity will develop to the target capacity level as opposed to the business-as-usual (BAU) level. SCA based policy formulation comes first, before deciding on the bundle of projects.

Social capacity development policy?

Chapter 1 provided the principles of SCD. But how can we formulate an effective aid policy to achieve SCD based on SCA? We will briefly explain three principles of SCD: comprehensiveness, ownership and sustainability, and the application of SCD to SCA.

Comprehensiveness

Comprehensiveness[1] implies that development and poverty reduction encompass all the elements of development – social, structural, human, governance, environmental, economic and financial factors (World Bank, 2006). Comprehensiveness in this SCD approach includes all the social actors, and their capacity can be developed under the relevant socioeconomic conditions and environmental performance. However, this comprehensiveness is not a mere aggregation of social actors and factors. It results in a synergy effect and creates social subsystems by the SCD approach. As discussed in Chapter 1, the system as a whole can bring about socioeconomic conditions and environmental quality in the SCEM.

The following is a summary of why SCA can achieve comprehensiveness: (1) social actors include the government, firms and citizens, which correspond to public, private with a profit orientation and private with a non-profit orientation, respectively; (2) factors include policy and measure, organizational resources, and knowledge and technology, which are the essentials for a policy cycle, and involve monitoring, analyzing and evaluating problems, and implementing sustainable policy; (3) SCA provides a framework, which includes the three social actors and three factors simultaneously; and (4) the SCEM approach results in a new subsystem, known as the social environmental management system (SEMS), which operates the SCEM.

The SCA based policy design assumes that the capacity of a country will not necessarily develop monotonically. It develops under the country's own socioeconomic conditions or environmental situation. As discussed in Chapter 1, in contrast to the developed countries, where environmental issues have shifted from being poverty related and industrial pollution related to consumption related, the developing countries are simultaneously affected by all the environmental issues. The required capacity in developing countries differs from that in developed countries. The SCA design is a tool that can be applied when socioeconomic conditions or the environmental situation differ. Then, comprehensiveness can be assured under the SCA based policy.

Ownership

Local ownership implies that development strategies must be formulated by developing (recipient) countries (that is, their governments and people) and must reflect their priorities, rather than the priorities of the donors (CIDA, 2000). 'Local' in local ownership includes the social actors who participate in the development.

The SCD process also ensures the ownership of developing countries. The process encourages the social actors' self-learning process, whereby they monitor, analyze and evaluate problems, and implement the sustainable policy. It is a self-learning process for the developing countries. Moreover, by nature, developing countries are diversified societies. The ownership itself and external assistance should also take this diversity into consideration. The ownership promised by SCD leads to a plural path of development in developing countries. We apply this idea to the SCD dimension. We can summarize why SCA can achieve ownership as follows: (1) an SCA based policy is based on the developing countries' leadership, which all the stakeholders are committed to; (2) all the stakeholders are asked to be responsible for development issues; (3) all the stakeholders can promote a self-learning process as they develop their capacity; (4) all the stakeholders participate in the self-learning process with regard to decision-making; and (5) the SCA process ensures a multifaceted path of development since developing countries by nature are a diversified society.

The purpose of the SCA based policy design is for enabling all the stakeholders to be engaged in the entire process, right from policy formulation to implementation and evaluation.[2] As the stakeholders get more rights for the development process, they develop their capacity by themselves.

Sustainability

Sustainability, an important concept in discussing capacity development, is more long-term. Sustainable development, which reveals the concept of sustainability in the development field, is defined as the development that meets the needs of the present without compromising the ability of future generations to meet their own needs (WCED, 1987).

The SCD process also tries to ensure sustainable development. SCD aims to receive the long-term benefit of capacity development in order to develop overall social capacity. The total system – comprising SCEM socioeconomic conditions and environmental quality – tries to develop under the country's preconditions, historical background and other structural conditions. This process pursues the construction of a sustainable future to bring about intergenerational equity in the country as a whole. This process, engaged in by all the stakeholders, can be an enormous source of sustainable development. Based on this assumption, we can summarize why SCA can achieve comprehensiveness.

The United Nations Conference on Environment and Development (UNCED) proclaims that sustainable development should be addressed as a special priority in developing countries and with regard to international actions (United Nations, 1992). One of the Millennium Development Goals adopted in the General Assembly in 2000 is to ensure environmental sustainability. This shows that sustainable development or sustainability is a high priority in the international development dialogue.

We can ensure sustainable development in the SCA framework because (1) SCA based policy aims to receive long-term benefits of capacity development in order to develop overall social capacity; (2) the SCA process places great value on developing countries' characteristics – for example, institutional structure or historical background; (3) SCA-based policy can predict the future development path to ensure intergenerational equity; (4) developed countries and developing countries are asked to engage jointly in the development process; and (5) SCA based policy aims at global sustainability as global citizens increase their capacity.

As discussed in Chapter 4 and on pp. 119–20, the capacity development process is affected by the institutional structure and historical background of the country, and the capacity does not increase in the short term. SCA based policy supports this robust social capacity and provides the appropriate projects in an adequate time frame. It covers all the resources, technology or knowledge to ensure sustainable development through SCD. Thus, an SCA based policy constantly supports the SCD process, following which sustainable development can be expected.

Furthermore, SCA based policy requires the participation of all the relevant social actors. Social actors are divided into stakeholders on the basis of their commitment to the issues. This point is similar to the principles regarding ownership. However, active participation guarantees the building of awareness and mutual discussion among actors. This enables the actors to widen their choices and opportunities to solve environmental issues and this contributes to sustainable social development.

Steps in policy design

This section provides step-by-step directions on how to develop a policy based on SCEM (see Matsuoka *et al.*, 2005). There are five steps involved in developing such a policy: scoping, selection of stakeholders, targeting, available resources for policy and the fulfilment of the five criteria for policy selection.

Step 1: scoping

The objective of scoping is to address issues comprehensively and grasp the conditions of each sector. To achieve this objective, we apply the concept of total system analysis. Figure 5.4 describes this concept.

The total system consists of four sectors: (1) socioeconomic conditions; (2) environmental quality; (3) SCEM; and (4) external factors. SCEM is characterized and driven by socioeconomic conditions. The interaction between SCEM and socioeconomic conditions governs environmental quality. Environmental quality and socioeconomic conditions have a mutual relationship. External factors affect all the sectors in terms of bilateral, multilateral and global aspects.

In this comprehensive concept of scoping, identifying environmental quality is a critical element in taking policy design forward. The environmental quality would define the level of SCEM and the stage;

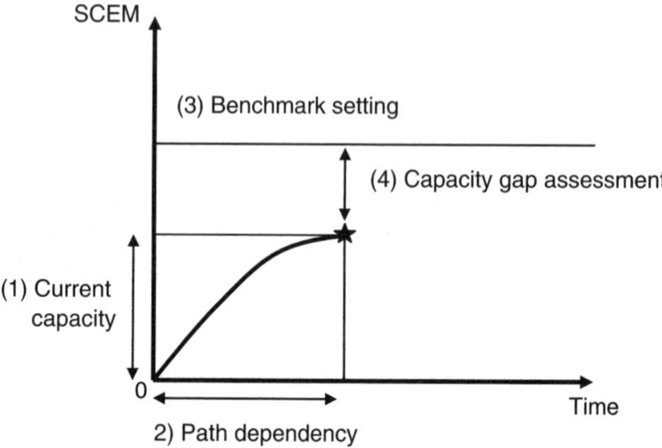

Figure 5.4 Process for assessing current SCEM

therefore, continuous and accurate data is required. This is why building and operating a monitoring system becomes a milestone in international environmental cooperation.

There do exist previous unfavourable examples, such as the 'Environmental Center Approach' (JICA, 2003), which was an environmental monitoring system. These examples were not sustainable or effective, and did not work well. There were two reasons why this happened. First, the level of capacity did not match the local conditions. Second, the local needs and their importance were not well traced. Therefore, it is significant to address issues comprehensively and capture the conditions related to capacity development for each sector. Environmental monitoring that defines the level of environmental quality is also critical. Table 5.1 outlines the key points of this subsection.

Table 5.1 Points of Step 1: scoping

1 Grasp conditions related to CD for each sector comprehensively in total system
2 Identify variables that define a level of environmental quality
3 Allocate environmental monitoring in scope of SCEM

Step 2: selection of stakeholders

The purpose of selecting stakeholders is to clarify the stakeholders that are being considered to address issues and conditionality. Stakeholders are the bases and key people selected for policy implementation; therefore, in this step, the people selected must be credited for policy design. Comprehensiveness must apply to all the stakeholders in this step.

Three actors – G (government), F (firm) and C (citizen) – constitute the framework by which to select stakeholders. A detailed list of the social actors included under G, F and C are different based on the direction of scoping. For example, G includes central and local government; F includes multinational companies, crown companies, industrial giants, medium-sized and small companies, and family businesses (see Kimbara and Kaneko, 2005a; 2005b). Although C encompasses an almost unlimited number of selected stakeholders, citizen groups such as NGOs are essential actors for SCEM development (see Yagishita, 2004; Cheng, 2005).

In Step 1, stakeholders must be appropriately selected to fit the conceptual space defined by the addressed issues and policies that can influence the selected stakeholders. For example, scales (regional, intermediate and local) must be considered in the selection of stakeholders when we focus on environmental issues such as air and water pollution, and soil contamination. Besides, effectiveness (or degree in operation) must be examined for the selected stakeholders. An environmental ministry is not the only body that influences environmental issues. Therefore, stakeholders must be selected over and above a careful analysis of social structure and addressed issues. Table 5.2 outlines the key points of this subsection.

Table 5.2   Points of Step 2: selection of stakeholders

1	Select stakeholders to fit G, F and C appropriately for setting the scope scale
2	Examine stakeholders' enforcement and role of development for SCEM
3	Examine a degree in operation for all stakeholders

Step 3: targeting

The objective of targeting is to set a goal determined by the process shown in Figure 5.4 (see also Figure 5.3). The current SCEM is estimated by actor–factor analysis, which reveals the level of social capacity by combining the results of both the actor and factor approaches (see Murakami and Matsuoka, 2006a; 2006b). The SCEM is also estimated by indicator development, which develops the indicators that carry summary information regarding SCEM (see Honda, 2004). Indicators describe the level of accumulated capacity based on fundamental variables and prescribe the social capacity derived from actor–factor analysis. How to measure SCEM and the capacity of social actors, and the level of capacity for each factor for each actor is a key issue (see Honda *et al.*, 2004; Tanaka *et al.*, 2005).

Path dependency and development stage analysis analyze the path through which the current level of social capacity can be reached (see Matsuoka *et al.* 2004). These analyses also identify a target level of capacity as well as information and conditions as assumptions to trace the path. Moreover, these analyses determine the development process of capacity by analyzing the development path and level of capacity for social actors (see Fujikura, 2004). Thus, indicator development, path dependency and development stage analysis are closely related, and enable us to study international comparison in terms of the development path of capacity and the level of capacity.

In the process of benchmark setting, we set a visible goal to move from the current status to the next stage, based on development stage analysis. Setting a numerical goal is difficult because SCEM expresses variability. Hence, setting a benchmark as a proxy that represents the development of SCEM is required. Benchmarks are achieved when social actors (G, F, C) and factors (P, R, K) work together in a holistic manner.

Capacity gap assessment specifies the gap between the benchmark and current SCEM. Figure 5.5 briefly represents the concept of this gap. The actor–factor matrix in Figure 5.5 provides a more detailed representation of the gaps for each social actor and factor during the proposal of concrete projects.

The significance of capacity gap examination based on the actor–factor matrix is to determine the developed capacity that has reached the

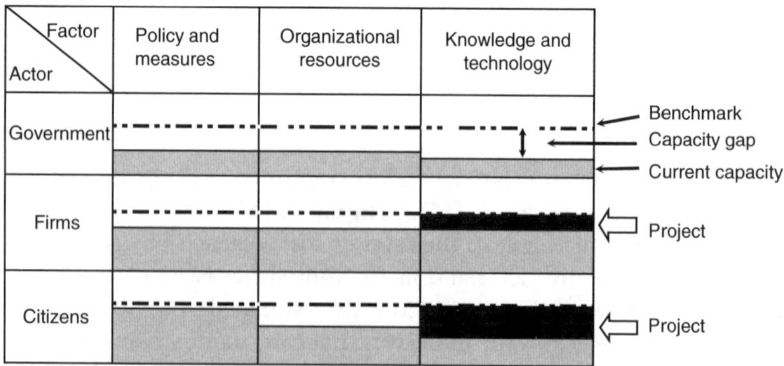

Figure 5.5   Actor–factor matrix model for assessing current SCEM

Table 5.3   Points of Step 3: targeting

1   Assess current capacity
2   Identify the development path of capacity and the status of stage
3   Set a benchmark for the next stage
4   Assess a capacity gap by benchmark

benchmark in advance (for example, the cell representing Citizen's Policy and Measures in Figure 5.5) so that the project can be detailed further. It also enables us to distinguish social actors who possess a high degree of expertise in operation. We can expect the endogenous development of such social actors and can provide resources to other weak actors. Table 5.3 the outlines key points of this subsection.

Step 4: available resources for policy

The objective of available resources for policy is to fill the capacity gap; filling the capacity gap is one objective of a project. This is not accomplished only with the help of ODA, FDI and so on. It should primarily be developed at the current capacity. We conceptualize current capacity in terms of resources; namely, policy and measures, organizational resources (human resources, funds and facilities), and knowledge and technology in the actor–factor matrix (Figure 5.5). Seeking available resources to fill the gap is a necessary precondition to develop an environmental policy based on SCEM. In terms of ownership, local resources should not only be sourced but should also be the foundation of ownership. Local resources include the developed capacity of social actors mentioned in the previous subsection. Social actors who have a high level of expertise in operation are a part of endogenous resources as well as formal and informal institutions or customs. Endogenous resources approve the sustainability of designing a policy.

We have no choice but to depend on external factors such as ODA for the budget and infrastructure. It is too difficult for local actors to obtain such high cost matters within a short period of time. A needs-based approach utilizing local instruments and resources from a recipient country leads to a sense of international cooperation. On the other hand, approaches involving the use of, for example, command and control[3] from the donor's side is preferred when immediate issues and benchmarks are in question. This approach would be more effective in avoiding the widening of irreversible problems and preventing irretrievable damages. If we were to disregard irreversibility, uncertainty, and externality – which are common characteristics of environmental problems – the costs of the solutions would increase drastically. Industrial countries that have experienced this negative reinforcement are aware of this logic and take it into account while formulating their policy. Developing countries, however, consider it as a secondary objective because they prioritize economic development. They have little sense of crisis, and it is still latent. Policy makers have to ensure simultaneous comprehensiveness, ownership, efficiency and sustainability.

The actor–factor matrix in Figure 5.5 is comprehensively applied to designing policy. We develop a project as a benchmark by seeking and listing endogenous resources (Figure 5.6). A project includes some cells in the actor–factor matrix. For example, building an environmental monitoring network and disclosing information is the task of G, while needs and opinions from C and cooperation from F are required to run the network. In short, in the actor-factor matrix, all projects need comprehensiveness. It is natural that an unnecessary cell can be found due to an adequate estimated capacity; however, we have to distinguish an estimated cell from an unrelated cell.

One of the projects that meets the benchmark should have alternatives for the social actors, based on the level of the factors. Let us suppose that 'setting effective environmental regulations and clarifying firms' is a benchmark. For this benchmark, the core actors are G and F, but the needs and opinions from C may be required to confirm effectiveness (see Matsumoto, 2004). Here, we can develop detailed projects to set

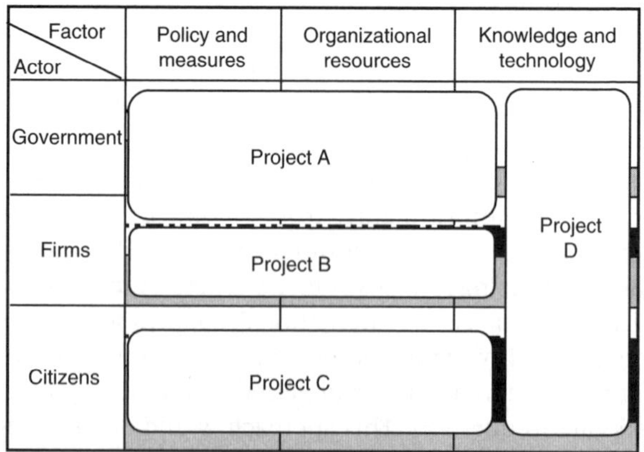

Figure 5.6 Project setting using the actor–factor matrix

Table 5.4   Points of Step 4: available resources for policy

1   Develop projects using actor–factor Matrix for gap filling of capacity
2   Seek and list available resources at the local level
3   Guarantee the ownership of projects by available resources
4   Note necessity of the dependence on external factor, by issue conditions

environmental standards for G and facilitate cooperation from F and C to fit their current capacity. We have to pay attention to social actors, factors and the extent of the capacity gap for this project development approach. Table 5.4 outlines the key points of this subsection.

Step 5: five criteria for policy selection

The objectives of the five criteria for policy selection are to develop several types of policies, which depend on the execution orders of the projects. Every policy must be characterized by comprehensiveness, ownership and sustainability during the administration of all the projects. Several types of policies can be implemented to fill the capacity gap, based on the integration of the projects. For example, we can design a policy to concentrate on approaching SCEM by a balanced, imbalanced and intermediate cell in the actor–factor matrix. Although those policies are optional and preferable, we should prioritize them on the basis of the five OECD–DAC criteria: effectiveness, efficiency, impact, sustainability and relevance (see Matsuoka and Honda, 2001). For example, if a policy is good with regard to impact but not efficiency or effectiveness, you can estimate the policy selection for its appropriateness. In addition, the policy for capacity development should be made a priority over others. Capacity development would be a comprehensive approach like the above haphazard appropriation. When we select policies taking into consideration the late comer's advantage, we can also apply some of the experiences of industrial countries to developing countries in order to find an effective sequence to formulate SCEM. Therefore, we have another way to evaluate policies – a more appropriate policy fitting local conditions, which is tailored by experiences from industrial countries.

The policies derived from the five steps are also examined based on the stage and path of development. We should consider the five OECD–DAC criteria and late departure related profit when trying to design more appropriate policies. Finally, when you select a policy, ensure that all the projects are consonant with the present capacity. In the next section, we

Table 5.5   Points of Step 5: 5 criteria for policy priority

1   Consider several types of policy development by execution orders of projects
2   Refer two ways for policies; OECD–DAC 5 criteria and late comer's advantage
3   All projects should be run wavy for arrival at benchmark on any policy selection

provide a tutorial of policy design. It may help practitioners to implement this sector in the field. Table 5.5 outlines the key points of this subsection.

Social capacity development for urban environmental policy

In this section, we design a policy for urban environmental management based on the SCA approach. Nowadays, needless to say, urban environmental problems are becoming serious. In this section, we considered City J as the subject city. City J is a megacity located in a developing country in East Asia. Most mega cities in developing countries like City J are plagued by problems of land use, traffic congestion, and environmental pollution.[3] We will design an urban environmental policy based on the five steps mentioned earlier: scoping, selection of stakeholders, targeting, resources available for policy design, and policy design. The important point is that City J carries out the process itself.

Scoping for urban environmental issues

First of all, we grasp an appropriate idea of the present situation regarding the total system in the field. According to the steps, in policy design (see pp.129–30), we have to have a comprehensive grasp on the conditions related to capacity development for each actor–factor and identify the proxies that define the level of progress of urbanization. Based on the entire system, an outcome of these researches is that City J is in the following situation (Figure 5.7).

SCEM

There are various players besides several levels of government; the players also include international companies and citizen associations. However,

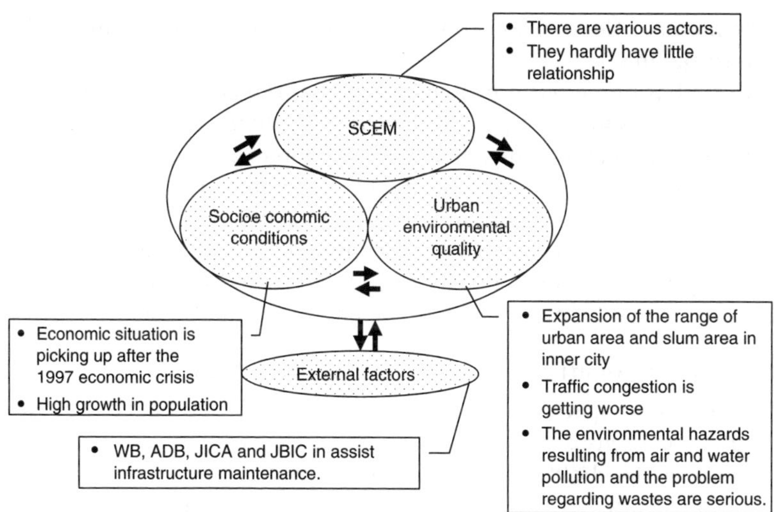

Figure 5.7   Total system analysis for City J

they barely have a relationship with each other. Further, we can identify formal and informal institutions that form an endemic community. There is devolution of power from central to local government.

Socioeconomic conditions

The economic situation is picking up after the 1997 economic crisis. City J's economy is advancing in the globalization of the economy. There is an increase in the growth of the population and there is high density population in the inner city. City J does not have a generalized social security system.

Urban environmental quality

City J is suffering from a lack of infrastructure maintenance. In a low-lying area along a river, an area is severely damaged due to floods. The urban and slum areas in the inner city have sprawled out. A sprawled urban area causes heavy traffic congestion, which results in serious air pollution.

External factors

The World Bank (WB), the Asia Development Bank (ADB), the Japan International Cooperation Agency (JICA) and the Japan Bank for

International Cooperation (JBIC) assist in developing basic and social infrastructures.

City J has numerous urban environmental problems: the problem of the suburban sprawl is especially serious owing to the rapid expansion of the population and highly disordered development. These urban issues exacerbate global warming and other environmental problems. Therefore, against such a background, we set an agenda for the development of a compact city.[4] A compact city is considered as one form of a sustainable city.[5] We could describe a compact city as follows: a high-density urban area that is anti-suburban in an environmentally friendly way.

Stakeholders of urban environmental management

To design a compact city, we have to formulate conceptual policies and facilitate consensus building among the major actors. In City J, there are various stakeholders to engage in consensus building (see Table 5.6). For instance, the government has several sectors such as central government, local government, affiliated companies and international organs. The central government sector consists of various departments and agencies: the Ministry of Urban Planning, Ministry of the Environment, Ministry of Economy and Trade and Industry, and the Ministry of Finance. We seek stakeholders vertically and horizontally. As with the government sector, the firm and citizen sectors also have various stakeholders.

When considering efficiency and leadership to implement policies, we could assume the city planning division of the local government as being the main actor. Because City J is developing, the government actor has autonomy in urban planning. Further, the other actors do not have much power and interest. The city planning division of local government is appropriate body to ensure that ownership, comprehensiveness and sustainability is attained in City J.

Targeting for the compact city

In this step, we set a visible benchmark as a goal to move from the current development stage to the next stage. Setting a benchmark as a proxy representing the development of the SCEM is required. We could consider the basic items necessary for a sustainable city as shown in Table 5.7.

Based on the actor–factor matrix shown in Table 5.7, we estimate the current SCEM. The critical minimum is categorized as two types: achievable and unachievable because it is easy to proceed in developing countries. Table 5.7 also shows City J's current SCEM based on the actor–factor analysis. Using the actor–factor matrix, we can grasp the capacity

Table 5.6 Stakeholders of urban environmental management in City J

Government	Central government	Ministry of Urban Planning	Ministry of Environment	Ministry of Economy and Trade and Industry	Ministry of Finance
	Local government	City Planning Division	Environmental Division	Commerce and industry division	Housing division
	Affiliated companies	Development of public cooperation	Corporation for Small and Medium-sized Enterprise	Housing cooperation	Environmental resource cooperation
	International organs	World Bank	ADB		
Firm	International major companies	Trading firms	General contractors		
	Small and medium-sized companies	Construction companies	Consultant companies		
	Associations	Chamber of Commerce and Industry	Labour union		
Citizen	International	NGOs	NPOs		
	National	NGOs	NPOs		
	Rural	NGOs	NPOs	Neighbourhood associations	

Table 5.7 City J's current SCEM and social capacity

	Policy and measures		Organizational resources		Knowledge and technology	
Government	Establishment of city planning and environmental law	A	Increase in urban planning budget	N	Conducting of surveys and analyses of city	A
	Formulation of urban plan	A	Implementation of public comments	N	Information disclosure and sharing	A
	Formulation of land use plan	A	Increase of urban planning countermeasure division staff	A	Conducting of case study on advanced city	N
	Improvement of public transportation in inner city	N	Recruitment and fostering of personnel responsible for urban planning	N	Relationships with research institutes	A
	Housing supply and improvement of residential environment	N			Implementation of staff training and education	N
	Development of control in the outskirts	A			Research on urban policies	N
	Policy coordination with other departments	N				
	Policy coordination with local government and central government	N				
	Round table discussions with firms and citizens	A				
	Tax deduction and subsidy system	N				
Firm	Observation of city planning law and regulations	N	Budget for land management	A	Implementation of staff training and education	N
	Utilization of the tax deduction and subsidy system	N			Network formation among firms	N
	Located in inner city	N			Introduction exchange	A

	Encouragement to use public transportation	N			
	Compliance with urban planning and fundamental plans	N			
	Participation in round table discussions	A			
	Policy recommendations on urban planning	N			
	Effective land use of company owned fields	A			
Citizen	Observation of city planning law and regulations	N	Organization of various events	A	Conducting of independent investigations	A
	Utilization of the tax deduction and subsidy system	N			Network formation among NPOs or NGOs or citizens associations	A
	Living in inner city	N			Information sharing and disclosure	N
	Formation of community based organizations	A			Information exchange	N
	Establishment of NPOs, NGOs	A				
	Active use of public transportation	N				
	Participation in round table discussions	N				
	Policy recommendations on urban planning	N				
	Reduction in use of private	N				

Note: A = achieved; N = not achieved.

gap. Let us focus our attention on Cell G_P. There are 10 items necessary for a compact city: (1) establishment of city planning and environmental law; (2) formulation of an urban plan; (3) formulation of a land use plan; (4) improvement of public transportation in the inner city; (5) housing supply and improvement of the residential environment; (6) development of control in the outskirts; (7) policy coordination with other departments; (8) policy coordination with central and local government; (9) round tables with firms and citizens; and (10) a tax deduction and subsidy system. With regard to Cell G_P, City J fulfills five out of nine items. When we view Cell C_K, there are four items necessary for a compact city J as follows: implementation of an independent investigation; network formation among NPOs, NGOs or citizen's associations; information disclosure and sharing; and information exchange. In Cell C_K, City J fulfills half the items. After analyzing all the cells, we find that City J fulfills 18 out of 45 items necessary to become a compact city.

After the results of a comprehensive analysis using the actor–factor matrix, we can say that City J has laws and regulations regarding urban environmental policy. However, these systems are not effective. Compared with the other actors, government has a high social capacity. When we consider the current social capacity of City J comprehensively, it seems that we have a critical cross-cutting issue with regard to efficient operation. Reflecting such a situation, City J is in the system making stage, which corresponds to the entry point position. We could set the following benchmark: good governance with cross-cutting working to achieve a sustainable city.

Available resources for urban environmental policy

Each cell has a greater or lesser capacity gap. For capacity gap filling, based on the actor–factor matrix, we seek and list available resources for sustainability at the local level (see Table 5.8). The projects must be developed by considering not only the capacity gap but also funds, capital, time, materials, information and the law. Furthermore, we have to ensure comprehensiveness, ownership and sustainability simultaneously. We will take Cell G_P as an example (see Table 5.9).

First, we assess the capacity gap for good governance with cross-cutting working to achieve a sustainable city. The gaps in Cell G_P are with regard to city planning law and formulation of an urban and land use plan. Although they have basic laws and regulations, there is a lack of rules for operation and resources. The central and local governments are not in cooperation with each other. Central government takes measures with a heavy hand; therefore, local government complains about this top-down

Table 5.8 Capacity gap and possible projects in City J

		Policy and measures	Organizational resources	Knowledge and technology
Government	Capacity gap	• Despite the establishment of a city planning law and formulation of urban plan and land use plan, the system is not working well • Central government and local government are not in cooperation with each other • There is no tax deduction and bounty system to encourage F and Cs initiative ↓	• Lack of necessary financial resources to improve infrastructure • Firms and citizens participate in some conferences, but their voice is not reflected • Staff are not suitable for the job ↓	• Government conducts surveys, but does not use the findings when formulating policies • Because of lack of an inadequate budget, they cannot conduct research on advanced area ↓
	Possible projects	(1) Simplification of the law and regulations to enhance control and observance (2) Formulation of rules, detailed regulations, and operation systems for effective implementation (3) Role-sharing arrangements among the stakeholders (4) Creation of a good relationship between central government and local government (5) New subsidy system for encouraging initiative	(1) Establishment of systems for their opinions to be reflected in politics (2) Development of specialist	(1) Conducting of surveys with research institutions (2) Facilitation of further research on urban environmental problems

Table 5.8 Continued

		Policy and measures	Organizational resources	Knowledge and technology
Firm	Capacity gap	• The roles are not observed to any significant degree ↓	• Some firms participate in the conference, but they are reluctant to join it ↓	• There is no activity for knowledge and technology ↓
	Possible projects	(1) Enhancement of the use of the subsidy system (2) Enhancement of information disclosure and sharing for firms	(1) Promotion of private-sector participation in urban policies	(1) Promotion of communication and information related activities
Citizen	Capacity gap	• Increase in the use of private vehicles • Some urban development projects ignored the community and residential environment ↓	• Different from civic organization • Lack of resources for civil activity • There are many NPOs and NGOs, but they have a shortage of qualified personnel and information and a low budget ↓	• Lack of resources for storing knowledge and technology ↓
	Possible projects	(1) Enhancement of control and observance (2) Community based residential environmental improvement	(1) Creation of a forum for civic organization (2) Establishment of new a subsidy system for civic activity	(1) Establishment of a new subsidy system for civic activity (2) Enhancement of communication with governments and firms

Table 5.9   Example of the process regarding policy resources

1 Assess the capacity gap for 'Good governance with cross-cutting working for a sustainable city'

- They have a city planning law and have formulated an urban and land use plan, but the system is not working well
- The central and local governments are not in cooperation with each other
- There is no tax deduction and bounty system to encourage F and Cs initiative
- They need a coordinator to create relationship between central government and local government

2 Seek and list available instruments and resources at the local level
There are minimum amounts of the following factors:

- Infrastructure
- Laws and regulations
- Staff
- Budget

3 Develop projects using the actor–factor matrix
For closing these capacity gaps, possible projects are as follows:

- Simplify the law and regulations to enhance control and observance
- Role sharing arrangement among the stakeholders
- Create a good relationship between central government and local government
- New subsidy system for encouraging initiative

They run the available instruments and resources with these projects. They need aid for international cooperation or government as follows:

- Enough budget to establish new subsidy system, etc.
- Assistance to enhance relationship among the stakeholders
- Advice and development social capacity

style. Despite this situation, there is no forum for discussion. There is the need for a coordinator to create a relationship between central and local government. There is no tax deduction and subsidy system to encourage firm and citizen initiatives. Hence, the urban environmental policies are not absorbed to a large extent by the firms and citizens. Next, we seek and list the available resources in City J. There is a strict budget and minimal infrastructure, laws and regulations, and staff. We could say that City J has minimum resources to maintain the current situation. Finally, we develop projects using the actor–factor matrix (see Table 5.10). To fill these capacity gaps, we set the following possible projects: (1) simplification of the law and regulations to enhance control and observance;

(2) formulation of rules, detailed regulations and operation systems so that implementation is effective; (3) role-sharing arrangements among the stakeholders; (4) establishment of a good relationship between central and local government; and (5) establishment of a new subsidy system to encourage initiative.

They utilize the available resources for these projects. However, they require some aid for international cooperation or government, for instance, adequate budgets to establish new subsidy systems, assistance to enhance relationships among the stakeholders and social capacity with regard to advice and development.

In all the cells, we have to develop and propose projects following this process. Using the actor–factor matrix, there are 19 projects to fill the capacity gap. In this process, it is necessary to thoroughly examine the necessity and suitability of the projects that are most important for sustainability. After the process, the appropriate policies are developed.

Urban environmental policy design

We have to design a policy that achieves the benchmark; this requires good governance with cross-cutting working to achieve a sustainable city. In this process, there are two important things. First, we have to consider interaction among actors and factors. Second, we have to give adequate thought to comprehensiveness, ownership and sustainability. When we consider City J's current SCEM, social capacity and capacity gap, it is essential to promote their effective application.

For good operation, we have to prioritize the 19 projects based on the OECD–DAC criteria; namely, effectiveness, efficiency, impact, sustainability and relevance. We could prioritize the following eight projects to promote the effective application for forming the compact city (see Table 5.11).

Cell G_P

(2) Formulation of rules, detailed regulations and operation systems for effective implementation;

Table 5.10   Actor–factor matrix

Actors/Factors	Policy and measures (P)	Organizational resources (R)	Knowledge and technology (K)
Government (G)	G_P	G_R	G_K
Firm (F)	F_P	F_R	F_K
Citizen (C)	C_P	C_R	C_K

Table 5.11   Policy design

	Policy and measures(P)	Organizational resources (R)	Knowledge and technology (K)
Government (G)	(2) Formulation of rules, detailed regulations and operation systems for effective implementation (4) Creation of a good relationship between central government and local government	(1) Establishment of systems for officers' opinions to be reflected in politics. (2) Development of specialist	(2) Enhancement of information disclosure and sharing for firms
Firm (F)			(1) Promotion of communication and information related activities
Citizen (C)	(2) Community based residential environment improvement		(2) Enhancement of communication with governments and firms

(4) Forming a good relationship with the central and local government.
Cell G_R
(1) Establishment of systems for officers' opinions to be reflected in politics;
(2) Development of a specialist.
Cell G_K
(2) Enhancement of information disclosure and sharing between firms.
Cell F_K
(1) Promotion of communication and information-related activities.
Cell C_P
(2) Improvement of community-based residential environment
Cell C_K
(2) Enhancement of communication with governments and firms.

As described before, in City J there is not a good relationship among the stakeholders. They have fundamental systems such as environmental law or urban planning law; however, there is no administrative instruction.

These problems inhibit the formation of a compact city. To promote the effective application of the policy, it is essential to strengthen the foundation of the administration and create a good relationship between all the actors. For this reason, we design policies with the eight prioritized projects.

As described so far, we can engage in policy design very simply and clearly using five steps. This measure is very useful, even in other fields.

Towards international environmental cooperation

A way forward for international environmental cooperation

Not only the SCEM theory described above but also most capacity development approaches were treated as one-country models and regarded international or regional cooperation as external factors. When the United Nations Development Programme (UNDP) developed a framework of capacity development, initially capacity was assumed to be that of one country. However, the aid replacement approach (i.e., the one-sided transfer of knowledge and technology from developed countries to developing countries) was insufficient to deal with the issues of international development assistance.

As times changed, the nature of the present environmental problem has also changed. Globalization has also affected environmental problems. This implies that the current environmental problems are not restricted to a single city or country. Thus, the internationalization of environmental problems is also an emerging issue. Global actions such as the Kyoto Protocol are representative of such issues. However, Carraro (2006) points out that the effective way to solve international environmental problems is not by global actions but by regional actions. He stresses that the effective solving of global problems is met with difficulty because adverse interests and concerns are increasing.

Furthermore, the relationship between trade and the environment is now discussed openly at the WTO and United Nations Conference on Trade and Development (UNCTAD) because of deepening globalization and regionalization. Closer economic integration raised the profile of environmental problems. Therefore, an ideal method of trade and economy and analysis of environmental cooperation is also necessary. The reason is that it is not believed that the deepening of economic integration automatically leads to the evolution of international environmental cooperation. Economic unification and environmental relations do not follow the same course.

The past analysis of SCEM focused on the resolution of the capacity gaps of each country. This is because the background of the model construction originated from and was based on brown issues from the 1970s in Japan when its environmental problems, caused by economic development, were solved individually by each country or city. For example SCEM contributed for solving environmental problems in Yokkaichi, where a city with the worst air contamination had been seriously affected during the rapid economic growth period.

However, the present difficulty is that some environmental problems cannot be solved by a single country; international environmental cooperation is required to solve these problems. Here, the need to expand the SCEM model to international environmental cooperation is apparent. As noted and discussed above, we assumed that international environmental cooperation is an external factor for the SCEM of each country. We now go on to suggest the possibility of model expansion from a one-nation model to a model of international environmental cooperation and regional integration. To achieve this result, the status and SCEM of the present east Asian environmental cooperation will be discussed. At the same time, we will evaluate the relationship between regional integration and regional environmental cooperation.

Revising east Asian environmental cooperation, based on the SCA

The purpose of this section is to illustrate the east Asian Environmental Cooperation with regard to SCEM and regional integration. The problem in East Asian countries is that most of them are developing countries with economic and cultural gaps. Moreover, there is extreme diversity among the East Asian countries compared with other regional integrations. As suggested by Kimura and Ando (2005a), the economic and political cooperation among east Asian countries increases every day. The framework of the Association of Southeast Asian Nations (ASEAN)[6] plus 3 (China, Japan, and Korea) plays a strong role in this deepening interdependence. However, the east Asian countries are so diverse that it is considerably difficult to bring about a simple integration.

In this case, to implement the SCEM of each country, international environmental cooperation can play a very strong role. International environmental cooperation is needed for the following three reasons. First, the failure of global environment management is apparent (Stiglitz, 2006). Second, there are transboundary problems between east Asian countries (IGES, 2005; Teranishi, 2005). Third, most east Asian countries do not have an adequate social capacity to solve and determine the problem and the policy (Honda, 2003).

The reality that east Asian countries are facing is that, in contrast to their needs, environmental cooperation has not yet borne fruit. This implies that there is something that makes international environmental cooperation in Asia difficult. There are two reasons for this: the first is, as Kimura and Ando (2005b) clearly points out, the differences in market integration – in other words, the integrated markets are not closed; the second reason is that, as Elliott (2003) claims, east Asian countries do not share and integrate common norms.

Regarding the economic aspect, interdependency between investment and trade is increasing. A characteristic of east Asia is that, as Kimura and Ando (2005a; 2005b) points out, the pattern of trade and investment is fairly original. Trade and investment concentrate on intermediate goods, which are never consummed within a single region. This implies that the cycle of production and consumption is not completed within a single region. The destination of final goods is mostly the USA and the EU.[7] These facts have two implications. One is that east Asia has to produce goods in accordance with the environmental standards of their final destination. The other is that the east Asian countries have to import pollution in return for the goods they export. This situation has double standards for the environment. Moreover, it makes it more difficult to organize regional standards or regional environmental policy.

As well as to the economic difficulties, east Asian countries also have difficulties with regard to norms.[8] As Elliott (2003) claims, the possibility of environmental cooperation based on sharing the concept of danger to human security would be the key to creating new environmental norms for east Asia. Although her idea is an ambitious and original one, in east Asia the widely accepted norm – Asian Value,[9] as Dr. Mahtier claimed, would contradict her idea and the norms would conflict with her views.

As described above, the economic and political situations of east Asian countries are different from those of the EU and NAFTA.[10] Compared with other regional integrations such as NAFTA and EU, (as past our studies show), the lack of international environmental cooperation is apparent. East Asian countries have a different pattern from that of the EU and NAFTA. The EU has the EU order to force each country to integrate EU law into their policy. A concrete environmental policy and standards, something that also has a positive effect when these countries are in the process of global negotiations, provides a basic SCEM to scope their issue, formulate their policy and implement it. The synergy effect on each social actor is, needless to say, important.

The case of NAFTA is also interesting. The environmental issue of NAFTA is a matter between Mexican industry and the trade unions of the USA. In particular, our previous studies show that pressure from the trade unions of the USA and environmental and labour related NGOs has a positive effect on improving SCEM in Mexico. This suggests that for developing countries, the pressure from final markets has a great impact on the implementation of environmental standards and laws. This implication is consistent with that of Carraro (2006) and his related research on the EU (see Table 5.12.)

In east Asia, the existing effective intergovernmental organizations are the United Nations Economic and Social Commission for Asia and the Pacific (UNESCAP)[11] and ASEAN. Taking an example from the ASEAN plus 3, the ASEAN sets strategies for ten common targets as environmental policy. These targets express the ASEAN commitments to the Johannesburg Plan of Implementation of the World Summit for Sustainable Development (WSSD) and provide the framework for ASEAN cooperation in ten priority areas: (1) global environmental issues; (2) land and forest fires and transboundary haze pollution; (3) coastal and marine environment; (4) sustainable forest management; (5) sustainable management of natural parks and protected areas; (6) freshwater resources; (7) public awareness and environmental education; (8) promotion of environmental management and governance; (9) urban environmental management and governance; and (10) sustainable monitoring and reporting and database harmonization (ASEAN Secretariat, 2004).

Table 5.12   Regional integration and environmental cooperation

	Norms	Markets	Environmental cooperation
EU	Strong	Inner-integrated	Strong SCEM
NAFTA	Weak	Inner-integrated	Pressure from C and F
ASIA	Weak	Outward-integrated	Very Weak and poor

From these 10 priority areas, it is the secretary of the ASEAN who has the capacity to raise issues – in other words, to recognize problem and scoping issues. However, with regard to the capacity to implement, ASEAN's capacity is still low. Examples of this capacity gap are AUS aid's comments on ASEAN environmental activity as follows: 'there is no evaluation and monitoring process'. Further, at the meeting of the Ministerial Conference on Environment and Development in Asia and the Pacific 2000, held at Kitakyushu, in Japan, from 31 August to 5 September 2000, the following conclusion was reached: 'the commonality between the ASEAN Strategic Plan of Action for the Environment and the regional action program 2001–2005 is the lack of funding sources. There is a need to bridge the gap between intention and action.'

Not only do east Asian countries have a very weak capacity to implement these targets effectively because of the lack of capital resources, human resources and coherent policy within the ASEAN, but also the lack of a coherent policy with the plus 3 is a big issue. This shows that social capacity for environmental management of this region is still low.

As has already been shown, capacity gaps requiring implementation of effective environmental policy should be recognized. However, it is difficult to identify a precise model of international environmental cooperation. From SCEM analysis, the required policy option and objects are as follows: to recognize and share the present and future environmental problems, to recognize and share the policy for international environmental problems, and to create a regional framework to solve international environmental problems.

An effective tool for international environmental cooperation from SCA is the indicator that was developed in Chapter 2. The indicator, following the logic of the Tobit model, shows the status of social capacity in the environmental management of one country as a comparable indicator within a region. This indicator is useful in two ways. One is to judge which country is at the entry point – the point at which foreign aid enters the country. The other is to categorize countries within a region as a group of donor countries or as a group of recipient countries. By this calculation, the most effective way to achieve aid coherence can be discussed.

Concluding remarks and sustainable implications

Applying the SCEM model to international environmental cooperation or regional integration, we can conclude that the present framework of international environmental cooperation and bilateral aid, which are based on G to G coordination, is far from adequate. Moreover, their synergy effects are very limited. Compared with that, what the SCEM model implicates for international environmental cooperation is that the new international actors, C and F, contribute an active role to create new linkage and synergy effects between C and C, F and F, etc. This also implies, although a very ambitious assumption, that the international environmental cooperation framework of the east Asian countries, based on the SCEM model, will destroy the conventional east Asian regime and international environmental cooperation theory.

The most important implication of the SCEM model for regional integration and international environment policy is that the actors of international environmental policy are not taken only by the government, but also by the citizens and firms. Further, the synergy effects between G, C, and F, as those between G and C, and C and F, as noted on pp. 136–48, are also the most important implications of the SCEM model compared with any other model.

The implications of the SCEM model for regional integration and international environmental cooperation have been loosely indicated. What is shown here is that east Asian regional integration should have a strong consensus to motivate and develop the framework. We also briefly discussed what is needed to create efficient environmental cooperation. The answer is integrated markets and integrated norms to make policy and aid more efficient. Without norms and integrated markets, international environmental cooperation should be recognized as a clear and present danger.

To achieve desirable results, all the actors have to increase their capacity. For example, for the government it is as important to bring about cooperation between Japan, Korea and China as it is to bring about cooperation with the ASEAN. For a civil society, making NGO and academic networks collect, share and discuss environmental and policy information is important. For firms, introducing ISO or CDM implies an environmentally friendly attitude.

The entry point has been discussed as a key point at which to organize and make concrete the aid policy. This idea is a very original point of the SCEM model. The internationalization of the SCEM model suggests the concentration of environmental aid in bilateral and multilateral aspects and seeks aid effectiveness. However, the needs of least developing countries (LDCs) should also be recognized in order to monitor and manage the environment and poverty. Therefore, increasing social capacity is also a current issue, since the transboundary environmental issues are significant matters that might directly impact poverty. This is another task for the SCEM model and its policy implications.

To solve these problems, the recognition of present issues is the most important task. Worst of all, compared with the economic and political trends, cooperation with regard to environmental issues is still modest. As a result, it can be said that these factors affect weak environmental cooperation among east Asian countries. There are also various gaps among east Asian countries. For instance, the difference in per capita GDP between Japan and the Lao Republic is more than 100 times. However, in addition to this type of economic gap, there also exist gaps with regard to cultural values and security. To overcome these gaps, the only and most effective way is to create social capacity with comprehensiveness, ownership and sustainability.

Notes

1 The World Bank applies this concept to the comprehensive development framework (CDF), which is one approach to guide development and poverty reduction, including the provision of external assistance (World Bank website).

2 This approach also considers the diversity of each developing country because we consider the SCD process based on each country's respective characteristics.

3 For instance, DKI Jakarta is one such typical megacity. In 2004, it had a population of 8,792,000. The population has risen sharply within the last 40 years. DKI Jakarta also has several urban environmental problems such as traffic congestion, air pollution and water pollution.

4 The relationship with the suburban area is very important. However, to simplify the designing of the policy, in this section we target the central city as a compact city.

5 There are various arguments that surround a sustainable city. For instance, Richard Rogers, who is a British architect, mentions the following features of a sustainable city: (1) just, (2) beautiful, (3) creative, (4) ecological, (5) convenient, (6) compact and polycentric, and (7) diverse.

6 The ASEAN's mandate is realizing the Hanoi Plan of Action in order to improve the state of the environment and ensure sustainable development in the ASEAN. The ASEAN secretary will publish the ASEAN environmental report in 2006.

7 The EU has the Environment Directorates-General (DGs) to implement environmental policy within a region. The Environment DG is one of 36 DGs and specialized services that constitute the European Commission. Its main role is to initiate and define new environmental legislation and to ensure that measures that have been agreed upon are actually put into practice in the member states. The Environment DG is based largely in Brussels and has around 550 staff.

8 Norms, in this article, are defined as common purposes in addressing environmental challenges and in influencing the pursuit of a regional community.

9 The former Malaysian prime minister, Mahathir bin Mohamad, was leading the charge. Other Asian leaders, notably Singapore's former prime minister, Lew Kwan Yew and current prime minister, Goh Chok Tong; China's president, Jiang Zemin and Premier Li Peng; and Indonesia's President Suharto have made similar accusations. They claim that in the Asian paradigm, values such as consensus and stability trump individual rights and freedom, and they argue that Western countries have no right to judge the humanitarian affairs of other nations.

10 The North American Free Trade Agreement (NAFTA) came into effect on 1 January 1994. The approach to the environmental side of the agreement – the North American Agreement on Environmental Cooperation (NAAEC) – was to create several new bureaucracies to monitor the environment, consider disputes and fund infrastructure improvements. The bureaucracies are the Commission for Environmental Cooperation (CEC), the Border Environmental Cooperation Commission (BECC) and the North American Development

Bank (NADB), both of which are under the US–Mexico Border Environmental Cooperation Agreement (BECA).

11   In the area of environment and sustainable development, the main objectives of UNESCAP are to promote regional and subregional cooperation for sustainable development, and to strengthen the national capacity of members and associate members in designing and implementing environmental and sustainable development policies and strategies that would enable them to maximize the benefits from globalization. Their best practices are as follows: the Kitakyushu Network for a Clean Environment, the Phnom Penh Regional Platform for Sustainable Development in Asia and the Pacific, State of the Environment in Asia and the Pacific 2000, and a publication entitled 'Water Conservation: A Guide to Promoting Public Awareness' (http://www.unescap.org/esd/aboutus/).

References

ASEAN Secretariat (2004) *Vientiane Action Programme*, 10th ASEAN Summit, Vientiane, November 2004 (http://www.aseansec.org/VAP-10th%20ASEAN%20 Summit.pdf).

Carraro, C. (2006) 'International Framework for Post Kyoto Protocol', Paper presented at Third World Congress of Environmental and Resources Economists, Kyoto International Conference Hall, Japan, July: 3–7.

Cheng, Y.Q. (2005) 'NGO's Activity and its Role in China', Discussion Paper Series (COE for Social Capacity Development for Environmental Management and International Cooperation, Hiroshima University), 2005-2: 8

CIDA (Canadian International Development Agency) (2000) 'Canada Making a Difference in the World: A Policy Statement on Strengthening Aid Effectiveness', Canadian International Development Agency, http://www. acdi-cida.gc.ca/INET/IMAGES.NSF/vLUImages/pdf/$file/SAE-ENG.pdf (accessed 25 August 2006).

Cybriwsky, R. and L.R. Ford (2001) 'City Profile Jakarta', *Cities*, Pergamon, 18, 3: 199–210.

Elliott, L. (2003) 'ASEAN and Environmental Cooperation: Norms, Interests and Identity', *The Pacific Review*, 16, 1: 29–52.

Fujikura, R. (2004) 'Role of Stakeholders in the Process of Japanese Successful Pollution during the 1960s and 1970s – Sulfur Oxide Emission Reductions in Industrial Cities', Discussion Paper Series, (COE for Social Capacity Development for Environmental Management and International Cooperation, Hiroshima University), 2004-2: 17 (in Japanese).

Hall, P. and C. Landry (1998) *Innovative and Sustainable Cities* (Dublin, Ireland: European Foundation for the Improvement of Living and Working Conditions).

Honda, N. (2003) 'Evaluating Social Capacity Development for Environmental Management in Developing Countries: Case Studies Institute of Developing Economics', Social Capacity Development for Environmental Management in Asia – Japan's Environmental Cooperation after Johannesburg Summit 2002, Shunji Matsuoka and Akifumi Kuchiki (eds), Tokyo, IDE-JETRO: 23–40.

Honda, N. (2004) 'The Role of Social Capacity for Environmental Management in Air Pollution Control: An Application to Three Pollution Problems in Japan', Discussion Paper Series (COE for Social Capacity Development for Environmental Management and International Cooperation, Hiroshima University), 2004-5: 13 (in Japanese).

Honda, N., S. Matsuoka and K. Tanaka (2004) 'Analysis of Causal Structure on Social Capacity Development for Environmental Management in Air Pollution Control in Japan', Discussion Paper Series (COE for Social Capacity Development for Environmental Management and International Cooperation, Hiroshima University), 2004-7: 16 (in Japanese).

IGES (Institute for Global Environmental Strategies) (2005), *IGES White Paper: Sustainable Asia 2005 and Beyond – In the Pursuit of Innovative Policies* (Tokyo: Urban Connections): 320.

Jenks, M. (2000) *Compact Cities: Sustainable Urban Forms in Developing Countries* (London: E&FN Spon): 356.

Jenks, M., E. Burton and K. Williams (1996) *The Compact City: A Sustainable Urban Form?* (London: E&FN Spon): 350.

JICA (Japan International Cooperation Agency) (2003) 'An Environmental Center Approach: Social Capacity Development for Environmental Management in Developing Countries and Environmental Cooperation', Evaluation Team on Environmental Cooperation, Japan Society International Development: 246.

Kimbara, T. and S. Kaneko (2004a) 'Possibility of Simultaneous Pursuit of Environmental and Economical Efficiency', Discussion Paper Series (COE for Social Capacity Development for Environmental Management and International Cooperation, Hiroshima University), 2004-6: 16 (in Japanese).

Kimbara, T. and S. Kaneko (2004b) 'Study on the Relations between Cooperate Environmental Performance and Environmental Management', Discussion Paper Series (COE for Social Capacity Development for Environmental Management and International Cooperation, Hiroshima University), 2004-8: 13 (in Japanese).

Kimura, F. and M. Ando (2005a) 'Two-dimensional Fragmentation in East Asia: Conceptual Framework and Empirics', *International Review of Economics and Finance*, 14, 3: 317–48.

Kimura, F. and M. Ando (2005b) 'The Economic Analysis of International Production/Distribution Networks in East Asia and Latin America: The Implication of Regional Trade Arrangements', *Business and Politics*, 7, 1, Art. 2.

Lavergne R. (2003) 'Local Ownership and Changing Relationships in Developing Cooperation', http://www.ccic.ca/e/docs/002_aid_2003-03_local_ownership_and_changing_relationships.pdf (accessed 30 August 2006).

Leitmann, J. (1999) *Sustaining Cities: Environmental Planning Management in Urban Design* (New York: McGraw-Hill): 412.

Matsuoka, S. and N. Honda (2001) 'ODA Project Evaluation and the OECD–DAC's Five Criteria', *Journal of International Development Studies*, Japan Society for International Development, 10: 49–70 (in Japanese).

Matsuoka, S., K. Murakami, N. Aoyama, Y. Takahashi and K. Tanaka (2005) 'Capacity Development and Social Capacity Assessment', Discussion Paper Series (COE for Social Capacity Development for Environmental Management and International Cooperation, Hiroshima University), 2005-4: 19.

Matsuoka, S., K. Murakami, N. Aoyama, Y. Takahashi and K. Tanaka (2005) 'Capacity Development and Social Capacity Development', Graduate School for International Development and Cooperation, Hiroshima University, Discussion Paper Series 2005-4, http://home.hiroshima-u.ac.jp/hicec/products/DP2005/DP2005-4.pdf

Matsuoka, S., S. Okada, K. Kido and N. Honda (2004) 'Development of Social Capacity for Environmental Management and Institutional Change', Discussion Paper Series (COE for Social Capacity Development for Environmental Management and International Cooperation, Hiroshima University) 2003-1: 25 (in Japanese).

Matsumoto, R. (2004) 'Development of Social Capacity for Environmental Management: The Case of Yokohama City', Discussion Paper Series (COE for Social Capacity Development for Environmental Management and International Cooperation, Hiroshima University), 2004-4: 15 (in Japanese).

Murakami, K. and S. Matsuoka (2006a) 'Evaluation of Social Capacity for Urban Air Quality Management', *Japanese Journal of Evaluation Studies*, Japan Evaluation Society, 6(1): 55–69 (in Japanese).

Murakami, K. and S. Matsuoka (2006b) 'An Empirical Study of the Methodology for Assessing Social Capacity: The Case of Urban Air Quality Management', *Japanese Journal of Evaluation Studies*, Japan Evaluation Society, 6(2): 43–57 (in Japanese).

Rogers, R. (1997) *Cities for a Small Planet* (London: *Faber & Faber*): 180.

Stephen, J. and G. Williams (2002) 'A Common Language for Managing Official Development Assistance: A Glossary of ODA Terms', http://www.opml.co.uk/publications/development_policy/a_common_languag.html   (accessed 25 August 2006).

Stiglitz, J. (2006) 'A New Agenda for Global Warming', *Economists, Voice*, 3, 7, Art. 3.

Tanaka, K., C. Ge, M. Suga and S. Matsuoka (2005) 'The Role of Environmental Management Capacity on Energy Efficiency: Evidence from China's Electricity Industry', Discussion Paper Series (COE for Social Capacity Development for Environmental Management and International Cooperation, Hiroshima University), 2005-1: 9.

Teranishi, S. (2005) 'The State of the Environment in Asia 2005–2006' (Tokyo: Springer-Verlag), (Japan Environmental Council, editor-in-chief with Awaji Takehisa): 395.

United Nations (1992) 'Report of the United Nations Conference on Environment and Development', Rio de Janeiro, 3–14 June 1992.

WCED (World Commission on Environment and Development) (1987) *Our Common Future* (Oxford: Oxford University Press): 400.

World Bank (2006) 'World Bank Strategies on Comprehensive Development Framework', http://www.worldbank.org/cdf/ (accessed 25 August 2006).

Yagishita, M. (2004) 'Evaluation of Nagoya Stakeholder Conference Aimed for the Realization of Environmentally Sound Material-Cycle Society based on Citizen's Participation', Discussion Paper Series (COE for Social Capacity Development for Environmental Management and International Cooperation, Hiroshima University), 2004-3: 10 (in Japanese).

Part II

Applications: Towards Effective Environmental Management

6
Social Capacity Development for the Reduction of Poverty

Akifumi Kuchiki

Introduction

This chapter aims to propose a flowchart approach to prioritize the policies of developing countries as objectively as possible by using data.

The three stages in the prioritizing of policies are divided into the following six steps:

Step 1 attaining the social subsistence level (education, health care and agriculture);

Step 2 macroeconomic stabilization;

Step 3 structural adjustment programmes;

Step 4 capacity building for economic growth (physical infrastructure and institutions);

Step 5 growth strategies;

Step 6 narrowing income inequalities and improving the industrial pollution related issues.

Prioritization of policies

If a budget is unlimited, it is best to implement all reforms at once as a package. However, every government needs to prioritize its reforms since policies demand changes in the government's budget allocations due to budget constraints. This section will propose a prototype model of one of the methods for prioritizing reforms. In this chapter, political stability is assumed to be a precondition for all policies and reforms. An a priori order of reforms will be given to the flowchart in Figure 6.1 based on the experiences of the development of developing countries as a prototype

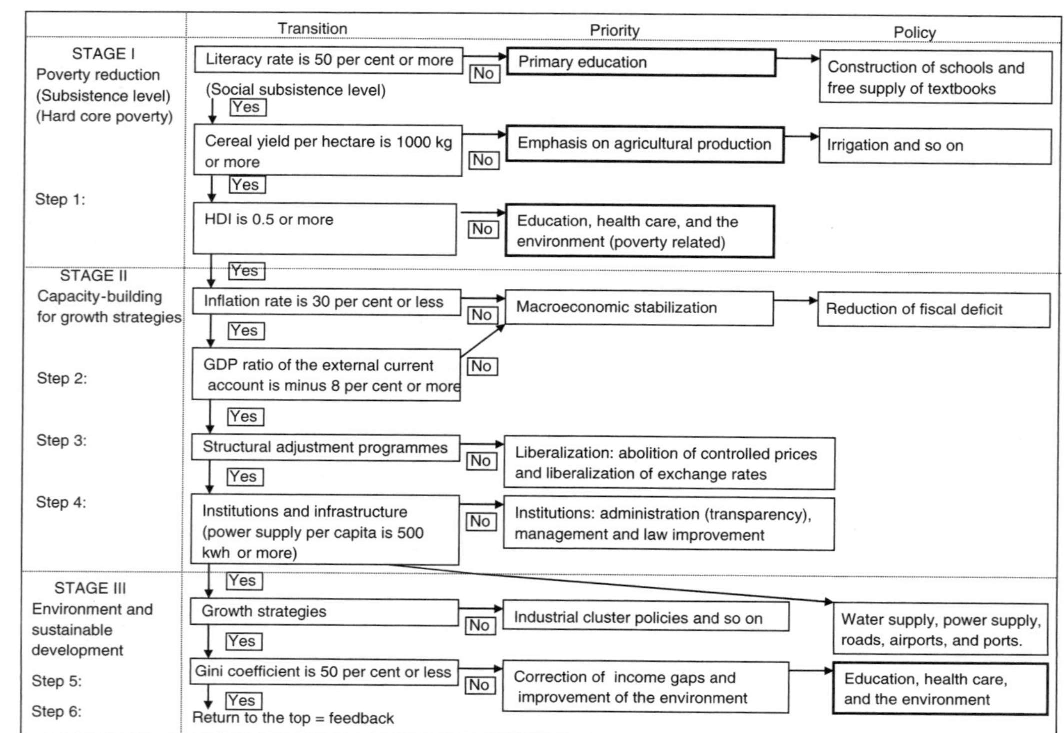

Figure 6.1 Flowchart approach of the prioritization method

model in the following six steps:

Step 1 Attain the social subsistence level of living-enhancement of the literacy rate (primary education, health care and environment) and agricultural reform (increase in agricultural productivity);
Step 2 Macroeconomic stabilization (stabilization of prices and stabilization of the balance of international payments) and stabilization of the financial system;
Step 3 Structural adjustment programmes (abolition of controlled prices, exchange rates liberalization and liberalization of interest rates), deregulation and the development of markets;
Step 4 Capacity-building for economic growth-improving physical infrastructure (roads, ports, airports, power and communication) and institutions that are closely related to the growth strategy;
Step 5 Growth strategy;
Step 6 Reducing income inequalities and improving the environment.

This chapter claims the following two points concerning the order of the six steps given in Figure 6.1. The first point is that for some countries, Step 1 (primary education and agricultural reforms to attain the social subsistence level of living) is required in the phase before both Step 2 (macroeconomic stabilization) and Step 3 (structural adjustment programmes). The second point is that Step 4 (improving physical infrastructure and institutional reforms) is a precondition for Step 5 (the economic growth strategy). The reasons for proposing the order of the following six steps of these reforms are explained below.

Step 1

As indicated in the United Nation Development Plan (UNDP), the ultimate purpose of economic growth is not growth itself but human development. The independent variables that determine the standard of human development are education and health. According to UNDP (2002), human beings aim to maximize social welfares including education and health care. Therefore, primary education is essential for achieving the social minimum standard of living. In a country having a literacy rate of less than 50 per cent, there is no doubt that the first priority will be given to primary education. For this reason, the social subsistence level appears at the top of the flowchart in Figure 6.1.

UNDP (2002) placed a 50 per cent critical value of Human Development Index (HDI) between the low and high levels in Table 6.1. In this chapter, we provide, a priori, a 50 per cent critical value of the literacy

Table 6.1 GDP per capita, life expectancy and adult literacy of countries with HDI lower than 0.5

	HDI	GDP per capita (PPP US$)	Life expectancy at birth	Adult literacy rate (Percentage ages 15 and above)
	2002	2002	2002	2002
Sierra Leone	0.273	520	34.3	36.0
Niger	0.292	800	46.0	17.1
Burkina Faso	0.302	1,100	45.8	12.8
Mali	0.326	930	48.5	19.0
Burundi	0.339	630	40.8	50.4
Guinea-Bissau	0.350	710	45.2	39.6
Mozambique	0.354	1,050	38.5	46.5
Ethiopia	0.359	780	45.5	41.5
Central African Republic	0.361	1,170	39.8	48.6
Congo, Dem. Rep. of the	0.365	650	41.4	62.7
Chad	0.379	1,020	44.7	45.8
Angola	0.381	2,130	40.1	42.0
Malawi	0.388	580	37.8	61.8
Zambia	0.389	840	32.7	79.9
Côte d'Ivoire	0.399	1,520	41.2	49.7
Tanzania, U. Rep. of	0.407	580	43.5	77.1
Benin	0.421	1,070	50.7	39.8
Guinea	0.425	2,100	48.9	41.0
Rwanda	0.431	1,270	38.9	69.2
Timor-Leste	0.436	–	49.3	58.6
Senegal	0.437	1,580	52.7	39.3
Eritrea	0.439	890	52.7	56.7
Gambia	0.452	1,690	53.9	37.8
Djibouti	0.454	1,990	45.8	65.5
Haiti	0.463	1,610	49.4	51.9
Mauritania	0.465	2,220	52.3	41.2
Nigeria	0.466	860	51.6	66.8
Madagascar	0.469	740	53.4	67.3
Yemen	0.482	870	59.8	49.0
Kenya	0.488	1,020	45.2	84.3
Zimbabwe	0.491	2,400	33.9	90.0
Uganda	0.493	1,390	45.7	68.9
Lesotho	0.493	2,420	36.3	81.4
Congo, Rep.	0.494	980	48.3	82.8
Togo	0.495	1,480	49.9	59.6
Pakistan	0.497	1,940	60.8	41.5

Source: UNDP, *Human Development Report 2002.*

rate and 10,000 kg critical value of cereal yield per hectare in Step 1. We also provide 0.5 critical value of HDI to take into consideration life expectancy, per capita income and so on. Note that this chapter will not discuss whether the 50 per cent critical value is reasonable.

Given below are the two reasons why agricultural reform is necessary for a country with a low rate of self-sufficiency in food production. First, according to some theory, it is necessary for a developing country to improve its efficiency in agricultural production in order to transform itself from a dual economy – where there is a gap in the productivity of agricultural and industrial sectors – into a market economy. Wages in a developing country cannot exceed the social subsistence level until its agricultural productivity exceeds the turning point of the theory. This is the so-called theory of the turning point proposed by Lewis (1955). Second, Myint (1965) explained the need for agricultural reform to increase agricultural productivity for economic development. Agriculture is important for the following three reasons: to supply an adequate amount of food, to provide labourers for nonagricultural (industrial) labour and to create demand for industrial products.

Governments should try to maximize self-sufficiency in food production in order to reduce the constraints of both a fiscal account and an external current account of an international balance of payments. A desert country cannot produce agricultural products. A country like Japan, which imports food and earns foreign currency by exporting manufactured goods, need not attain self-sufficiency in food production. However, most of the developing countries face both fiscal and external current account deficits that cause macroeconomic instability. This step can be skipped if a country does not encounter budgetary and foreign reserve problems without increasing the self-sufficiency in food production. We also adopt Myint's hypothesis and set the agricultural reform to Step 1 of Stage I in the flowchart on prioritizing policies in Figure 6.1. It is observed that the order and critical values of the reforms in Figure 6.1 should be justified by further discussing whether they are reasonable.

It can be demonstrated that the economic growth of developing countries accelerated due to the increase in their agricultural productivity (Kuchiki, 1994). After the 1980s, Asian countries such as China, Laos, and Vietnam implemented agricultural reforms prior to the structural adjustment programmes of Step 2. For instance, the Chinese government introduced a farmer production contract system at almost the same time as it introduced the open-door policies of liberalization in 1979 in order to give incentives to farmers. As a result, food production was increased dramatically; this reduced the amount of direct and indirect subsidies to agriculture, thereby contributing to the achievement of macroeconomic stabilization by restoring its fiscal discipline. The policy of the farmer production contract system was successfully applied to other Asian socialist countries during the transition period from a planned economy to a market economy. For instance, a similar incentive policy was implemented for the farmers in Vietnam and Laos. We could find a counter-example in Russia showing that the structural adjustment programmes had an adverse effect at least in the 1990s, because agricultural reform might not have been implemented sufficiently before Step 2.

Thailand adopted an agricultural policy in the 1950s and 1960s to increase its productivity of rice. Rice was important for Thailand in the post-Second World War period to earn foreign currency, since taxes on rice exports constituted 32 per cent of the budget revenue in 1953 (MacIntyre, 2001). Malaysia adopted the policy of shifting rubber production to oil palm production in the 1970s, and laid emphasis on the fostering of natural resource based industries such as petroleum, tin and palm oil to improve the value added in the 1980s. The Ministry of Primary Industries was established in 1972 to play an important role in the national economy of the 1970s through its contribution of some 70 per cent of the total export (ESCAP, 1999). Japan adopted an agricultural policy to earn increased foreign currency by exporting silk and tea in the 1890s (Kawamura, 2002). We assume that agricultural reform is a key to attaining macroeconomic stability in many countries.

According to Bai and Imura (2000), there exist three types of environmental issues from the perspective of economic levels; poverty related issues, industrial pollution related issues, and consumption related issues. The poverty related issues, such as access to safe drinking water and sanitation, should be solved in Step 1. The industrial pollution related issues should be raised in Step 6.

Step 2

From the 1980s onwards, the World Bank and the IMF began to implement structural adjustment programmes as a package of reforms. We will explain the reasons for providing the above-mentioned order of Steps 2 and 3. The IMF imposed a conditionality of macroeconomic stabilization on recipient countries as a precondition for the loans granted by the World Bank. The World Bank then imposed the conditionality of implementing economic liberalization in both the 1980s and the 1990s through structural adjustment programmes. We believe that these two steps are preconditions for introducing the economic growth strategy.

Step 3

According to Burnside and Dollar (2001), the order was successful in the case of many developing countries in the latter half of the 1980s and the first half of the 1990s, and it showed that aid is effective in developing countries with sound policies of macroeconomic stabilization. This is the reason why we adopt the order of Steps 2 and 3. In our opinion, the order of reforms should be as objective as possible.

Structural adjustment programmes include many reforms, such as trade liberalization, abolition of controlled prices and capital liberalization. The World Bank requested recipient countries to implement these reforms concurrently as one package, particularly in the 1980s and in the first half of the 1990s. Based on the experiences of the 1997 Asian currency crisis, some economists concluded that a rapid inflow of short-term capital to countries such as Thailand and Malaysia generated bubble economies due to the liberalization of movements of short-term capital, such that an outflow of the capital caused the Asian currency crisis (Furman and Stiglitz, 1998). They insisted that short-term capital liberalization in developing countries be restricted at an early stage of development. In future, there will be a need to reexamine whether the liberalization of long-term capital in Step 3 of structural adjustment programmes (Ito, 2001) needs to be prioritized over that of short-term capital.

Step 4

The physical infrastructure and institutions required for the growth strategy should be facilitated in Step 4. For example, the growth strategy in the manufacturing sector is closely related to how sufficiently the transport sector addresses issues concerning roads, ports and airports.

Physical infrastructure is divided into the following two aspects: what relates to basic human needs and what relates to the growth strategy. An example of the former aspect is water services for subsistence of human life; that of the latter is ports and highways for effective implementation of the growth strategy. Thus, in Step 4, it is important to focus on the physical infrastructure required for the growth strategy. We must facilitate the physical infrastructure that relates to the growth strategy in Step 5. In the case of ASEAN countries in the 1980s, the physical infrastructure – such as ports, electricity and roads – was limited to their industrial zones in order to invite foreign direct investment (FDI) (Kuchiki and Yamada, 1997). The physical infrastructure can be termed as the economic infrastructure.

There are many institutions to be reformed in each country. Section 4 presents an analysis on the reforms of Laos, Vietnam and China. Research on institutions is crucial to the successful implementation of reforms. Usually, each step has its own institutional reforms. Here, we limit our focus of institutional reforms in Step 4 to reducing corruption and preparing an investment climate for the growth strategy in Step 5.

Step 5

There exists an argument with regard to whether the selective interventions of government are needed to implement economic growth (World Bank, 1993). It is assumed that the growth strategy can be adopted if the above-mentioned preconditions hold. In short, next to Step 1, we prioritize the remaining steps as follows: Step 2 (stabilization of macroeconomy); Step 3 (structural adjustment programmes); Step 4 (improvement of infrastructure and institutions); and Step 5 (the economic growth strategy).

On completion of Step 4, governments can play an important role in implementing Step 5. The strategies can be illustrated by developing a private sector and promoting the introduction of FDI.

It is especially difficult to plan for industrialization in Step 5. The growth strategy is dependent on the stage of economic development in a country. It will be shown that identification of 'economic agents' and establishing of 'institutions' are necessary for the growth strategy in each developing country. The strategies of Morocco are illustrated in Figure 6.2 and explained below.

1   Consider a country whose key products are primary commodities. Then, its strategies may take forms of agro-based industrialization, resource based industrialization and export of primary commodities. This growth strategy was successfully adopted for cocoa, palm oil and petroleum in Malaysia in the early 1980s, subsequent to export led growth in the electronics industry in the latter half of the 1980s.

2   The following are candidates of economic agents that implement the growth strategy: (i) traditional industries; (ii) state-owned enterprises; (iii) export industries of nontraditional goods; and (iv) small and medium-sized enterprises.

3   The introduction of foreign direct investment (FDI) in east Asia was the most popular and successful strategy adopted all over the world, particularly after the 1980s. There exist two types of industries under FDI; that is, import substitution industries and export-oriented

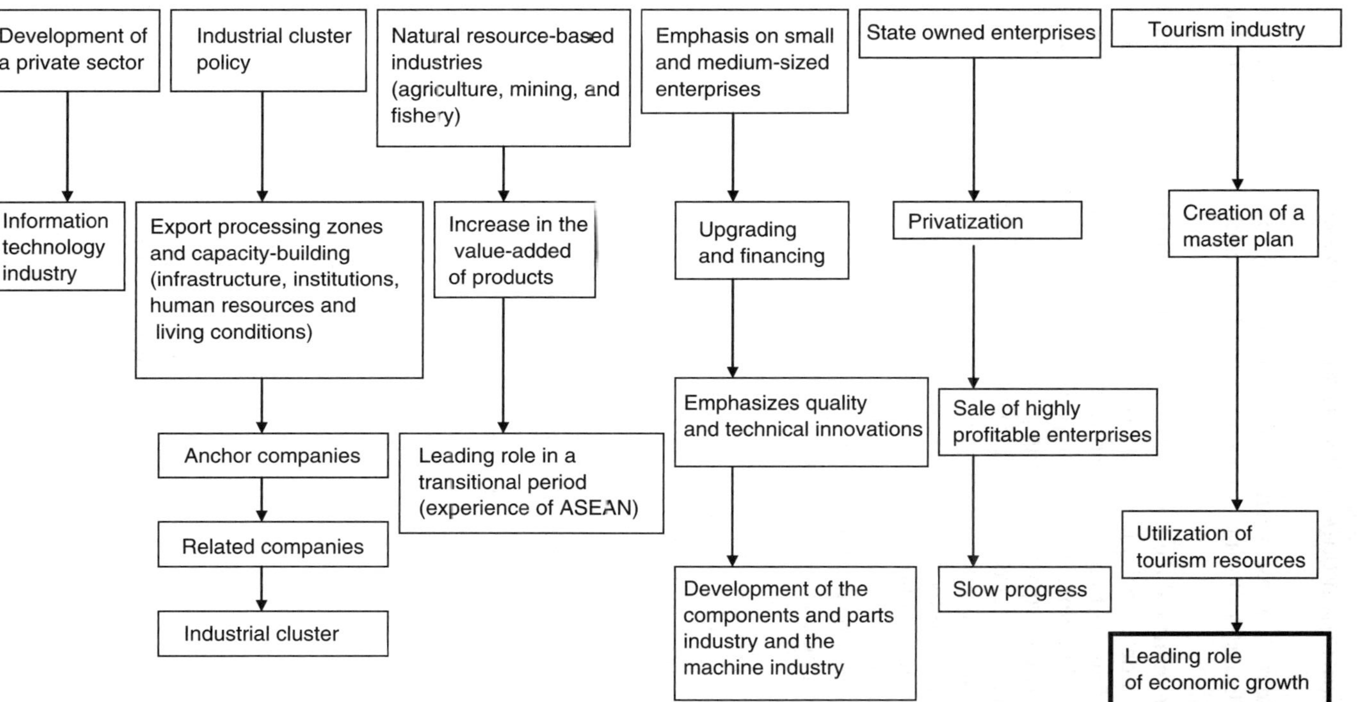

Figure 6.2 Growth strategies of Morocco

industries. The east Asian countries successfully introduced export oriented industries in the latter half of the 1980s and the first half of the 1990s.

4   The industrial cluster policy is a typical growth strategy adopted for regional development in Asia. Kuchiki (2005) referred to this strategy as the flowchart approach, which is a flow of export processing zones, capacity-building, anchor companies and their related companies. In this chapter, capacity-building means to facilitate the physical infrastructure, reform institutions, develop human resources and satisfy living conditions. A regional development strategy involves the establishing of export processing zones and/or industrial zones with preferential treatments, such as tax reductions or exemptions, in order to invite FDI. The export led growth strategy under FDI was adopted in east Asia. This was realized by developing export processing zones that contributed to the economic growth in east Asia until the 1997 Asian currency crisis occurred.

5   If a country has the resources for tourism, then the tourism industry can be used to earn foreign currencies to import raw materials for the manufacturing industry.

Step 6

After a country attains high rates of economic growth in Step 5, growth is sometimes accompanied by income inequalities between the rich and the poor, and environmental pollution. Therefore, we must adopt policies to solve the problems that exist in Step 6.

On one hand, Kuznets (1955) proposed a hypothesis of the inverted U-curve; that is, the conjecture that inequality would first rise and then fall with economic development. On the other hand, Deininger and Squire (1996) did not find any empirical support for Kuznets hypothesis. In this chapter, we assume Kuznets' hypothesis such that the growth strategy is adopted in Step 5 and the narrowing of the gap occurs in Step 6. A country usually becomes politically unstable when income inequalities become wider. The inequalities might bring about social unrest. This step cannot be avoided when a country faces political instability due to the gap. The environmental pollution problem becomes serious in Step 6 and is solved according to Matsuoka and Kuchiki (2003).

Regional characteristics

In this section, the flowchart method is applied to the developing countries of 1993, 1994, 1995, 1998, 1999 and 2000 in order to demonstrate its effectiveness in determining regional characteristics. This effort will identify the policies that governments should aim to prioritize.

The characteristics of each region in the world will be presented through the use of the data provided in Tables 6.1–6.9. The characteristics could be classified in the following manner: for sub-Saharan African countries, the first priority was given to Step 1 (primary education). For South America, Step 1 (improving self-sufficiency in food production) was emphasized. For east Europe and Central Asia, Step 2 (stabilization of inflation) was necessary from 1990 to 1995. Countries with comparatively high incomes in the Middle East adopt Step 2 (the external balances of a current account). South America is under the pressure of Step 6 (correction of income inequalities). These conclusions can be derived using our flowchart method together with the data provided in Tables 6.1–6.9.

Step 1

If the literacy rate is 50 per cent or less, the first priority is given to primary education. It is necessary to improve the literacy rates in countries that have an income per capita of 600 dollars or less, where the average life of population is 50 years or less; for example, the sub-Saharan African countries such as Mozambique, Ethiopia, Burundi, Chad, Mali, Gambia and Benin (Table 6.1). Although we give a priori a 50 per cent value of a criterion for deciding, at this step, whether or not to proceed to the next step, there are other values of the criterion that we can adopt. Therefore, we propose a prototype model to further discuss whether the value of 50 per cent is reasonable.

Improving self-sufficiency in food production is a prerequisite for macroeconomic stabilization policy. In order for a country to produce agricultural products with low productivity, its agricultural productivity must be improved; for example, a country whose cereal yield per hectare is less than 1000 kg belongs to this category. The countries belonging to this category are Algeria, Morocco and Libya in North Africa; and Botswana, Central Africa and Sudan in the sub-Saharan African countries. Many sub-Saharan African countries were placed in this category in both 1995 and 2000 (Table 6.2).

Step 2

All the countries should achieve macroeconomic stability as a prerequisite for the introduction of the growth strategy. Macroeconomic stability implies the stabilization of inflation, an external balance of a current account of an international balance of payments and a balance of a fiscal account. Countries requiring stabilization of the balance of the

Table 6.2 Cereal yield

	(Kg per hectare)	
	1995	2000
Algeria PRD	397	619
Botswana R	206	203
Burkina Faso	782	887
Cape Verde R	262	312
Central Africa R	868	1111
Chad R	586	626
Congo DR	826	774
Congo R	826	686
State of Eritrea	516	703
Grenada	983	1000
Haiti	921	913
Iraq R	800	284
Jordan	997	891
Kazakhstan R	578	1060
Kingdom of Lesotho	774	988
Libya	677	727
Mali	806	1170
Mauritania	771	1010
Mongolia	737	956
Kingdom of Morocco	446	471
Mozambique R	652	948
Namibia R	194	473
Niger R	291	368
Senegal R	873	745
Somalia DR	471	488
Sudan R	424	529
Tajikistan R	961	1290
Togo R	790	953
Vanuatu R	538	538
Lao RPD	2492	3184
Vietnam SR	3569	4048
China PR	4663	4735

Notes: 1 Countries of cereal yield of less than 1000 kg in 1995 except Laos, Vietnam and
China.

   2 The data of Morocco, Laos, China and Vietnam are used for Table 6.11.

Source: World Bank (2001).

current account were Jordan, Lebanon, Oman and Saudi Arabia among
the Middle Eastern countries (Table 6.3). Many east European and
Central Asian countries had to stabilize inflation in the first half of
the 1990s, including Albania, Azerbaijan, Belarus, Croatia, Estonia,

Table 6.3   Current account balance

	Percentage of GDP 1993–95		Percentage of GDP 1999
Congo, Rep.	−30.9	Armenia	−16.6
Guinea-Bissau	−21.4	Azerbaijan	−24.5
Honduras	−8.0	Bahamas	−14.8
Hungary	−8.9	Belize	−8.9
Jordan	−16.0	Bhutan	−19.7
Laos PDR	−13.7	Burkina Faso	−12.1
Lebanon	−44.7	Cape Verde	−17.3
Madagascar	−8.5	Chad	−10.2
Malawi	−19.5	Dominica	−9.4
Mali	−8.6	Eritrea	−43.4
Mauritania	−9.3	Gambia	−11.7
Nicaragua	−37.4	Ghana	−9.8
Omen	−9.2	Kyrgyz	−14.8
Paraguay	−14.9	Latvia	−9.7
Saudi Arabia	−9.9	Lebanon	−34.1
Tanzania	−19.4	Lesotho	−23.0
Togo	−9.0	Maldives	−17.8
Uganda	−9.1	Nicaragua	−26.5
Vietnam	−8.6	Niger	−8.6
China	−0.4	Panama	−14.3
Morocco	−3.3	Poland	−8.0
		Portugal	−8.5
		Seychelles	−16.8
		St Kitts	−36.4
		St.Lucia	−10.8
		St Vincent	16.2
		Togo	−9.0
		Turkmenistan	−17.2
		Uganda	−11.6
		Laos RPD	−6.1
		China	2.1
		Vietnam	−0.2

Note: Countries of current account balance of less than −8%, except China, Morocco, Laos and Vietnam.

Source: World Bank (2001).

Latvia, Lithuania, Macedonia, Poland, Rumania, Slovakia and Ukraine (Table 6.4). Countries with relatively high incomes are required to stabilize the balance of fiscal accounts, including Finland, Greece, Italy, Russia and Sweden (Table 6.5).

Table 6.4   Inflation

	Consumer prices % 1990–95		1999
Albania	64.2	Angola	559.7
Azerbaijan	1005.5	Belarus	322.0
Belarus	1247.2	Ecuador	61.9
Brazil	1044.8	Equatorial Guinea	38.3
Croatia	328.0	Kyrgyz	37.5
Ecuador	40.0	Laos PRD	126.3
Estonia	52.8	Malawi	42.1
Guinea-Bissau	45.1	Russia	64.6
Jamaica	39.9	Suriname	180.1
Latvia	83.1	Turkey	56.2
Lebanon	36.8	Uzbekistan	43.9
Lithuania	124.1	Zimbabwe	48.1
Macedonia, FYR	397.9	Morocco	0.9
Mozambique	47.8	China	−1.4
Nicaragua	85.2	Vietnam	5.5
Nigeria	49.1		
Peru	69.0		
Poland	41.5		
Romania	151.5		
Russian Federation	381.6		
Sierra Leone	41.9		
Slovenia	62.1		
Sudan	114.3		
Turkey	79.3		
Ukraine	1180.4		
Uruguay	59.3		
Venezuela	43.8		
Zaire	3558.3		
Zambia	112.5		
China	11.4		
Morocco	5.8		

Note: Countries of inflation of more than 30 per cent except China, Morocco and Vietnam.

Source: World Bank (2001).

Step 4

One of the prerequisites for growth strategy is improvement of the physical infrastructure and institutions. Among these prerequisites, first priority is given to the improvement of the power supply. Some Asian countries requiring improvements to their power supply were India, Indonesia, Pakistan, the Philippines, Sri Lanka and Vietnam (Table 6.6).

Table 6.5 Overall budget deficit (including grants)

	Percentage of GDP 1994		Percentage of GDP 1999
Ethiopia	−8.5	Columbia	−7.0
Finland	−13.5	Guinea	−7.1
Greece	−15.7	Lithuania	−7.0
Italy	−10.5	Maldives	−9.3
Oman	−11.2	Mongolia	−10.4
Russian Federation	−10.5	Sri Lanka	−7.4
Sri Lanka	−8.5	St. Vincent	−7.3
Sweden	−12.8	Thailand	−11.1
Yemen, Rep.	−17.3	Turkey	−13.0
China	−1.8	Morocco	−1.4
Morocco	−1.4	China	−2.9
		Laos RPD	−4.0
		Vietnam	−2.8

Note: Countries of overall budget deficit of GDP of less than −8.0 per cent in 1994, less than −7.0 per cent in 1999 except China, Morocco, Laos and Vietnam.

Sources: World Bank (2001); data for Lao RPD and Vietnam are from *Asian Development Outlook 2001* from the Asian Development Bank.

It is observed that the power supply is provided by the private sector, not the public sector, if it is expected to be profitable. The role of governments is to prepare an investment climate for both domestic and foreign investors in the private sector in order to encourage investment in electricity services.

Step 6

Countries with large income inequalities in South America, such as Brazil, Chile, Colombia, Dominica, Guatemala, Honduras, Mexico, Nicaragua, Panama and Venezuela, have been included in Table 6.7.

In this chapter, we will not discuss Steps 3 and 5 since we do not have the data related to those steps.

Applying the prioritization method to Laos, Vietnam and China

Laos

The government of Laos prioritized the improvement of the infrastructure, state run enterprises reform, the introduction of FDI, development of small and medium-sized enterprises, and development of the

Table 6.6   Energy production 1994 and 1998

	Kwh per capita 1994		Kwh per capita 1998
Bolivia	390	Bangladesh	102
Cameroon	212	Bolivia	466
Congo	172	Cameroon	229
Côte d'Ivoire	170	Congo, Dem. Rep.	117
Ghana	368	Congo, Rep.	123
Guatemala	306	Ghana	394
Honduras	464	Guatemala	412
India	423	India	504
Indonesia	281	Indonesia	382
Kenya	136	Kenya	166
Morocco	426	Morocco	508
Nicaragua	398	Mozambique	404
Nigeria	144	Myanmar	102
Pakistan	463	Nicaragua	472
Philippines	404	Nigeria	130
Senegal	121	Senegal	142
Sri Lanka	246	Sri Lanka	302
Vietnam	170	Vietnam	283
Yemen, Rep.	146	Yemen	151
China	778	Laos RPD	n.a.
		China	938

Note: Countries of energy production between 100 kwh and 500 kwh except China, India and Morocco.

Source: World Bank (2001).

traditional industry, tourism, the processing industry of primary goods and farming.

Now, we will apply the flowchart method to the case of Laos using Tables 6.1–6.10 and the summary of Table 6.11. It is clear that since the literacy rate in Laos is 48 per cent (that is, its illiteracy rate is 52 per cent) in 1999, it should give priority to primary education. The slash-and-burn farming method practiced in Laos is currently preventing the diffusion of education. The children of this country cannot attend school every day since their parents keep moving from one place to another. Therefore, it is important to modify the farming method before modifying educational reform. Construction of an infrastructure, such as roads, is necessary for its growth strategy. The cereal yield per hectare of Laos will satisfy the criterion for agricultural production of more than 1000 kg per hectare. However, it is important for the economy to attain

Table 6.7   Distribution of income or consumption (various years)

	Gini index	Year
Brazil	60.0	1996
Central African Republic	61.3	1993
Chile	56.5	1994
Colombia	57.1	1996
El Salvador	52.3	1996
Guatemala	59.6	1989
Honduras	52.7	1996
Lesotho	56.0	1986–87
Mali	50.5	1994
Mexico	53.7	1995
Nicaragua	50.3	1993
Niger	50.5	1995
Nigeria	50.6	1996–97
Papua New Guinea	50.9	1996
Paraguay	59.1	1995
Sierra Leone	62.9	1989
South Africa	59.3	1993–94
Morocco	39.5	1998–99
Laos RPD	30.4	1992
Vietnam	36.1	1998
China	40.3	1998

Note: Countries of distribution of income of more than 50 except Morocco, Laos, Vietnam and China.

Source: World Bank (2001).

macroeconomic stability since its inflation rate is higher than our criterion of 30 per cent. It is observed that the ratios of both its external current account and the overall budget deficit to GDP exceed our criteria by more than 8 per cent. It is necessary to have institutional capacity-building in Step 4 to improve transparency and reduce corruption. As a first step under the current political system, it is important to conduct research on the problems related to institutional capacity-building. Thus, it can be stated that Laos should not place significant emphasis on the improvement of the growth strategy before that of educational reform, the attainment of macroeconomic stability and institutional capacity-building in Steps 1, 2 and 3.

Vietnam

The policies of Vietnam can focus on growth strategy, which is different from the policies of Laos since it satisfies the criteria of the

Table 6.8 External debt (percentage of GNI) and debt service (percentage of exports), selected countries

	External debt 1995	**Debt service 1995**
Mozambique	443.6	35.3
Sierra Leone	159.7	60.3
Guinea-Bissau	353.7	66.9
Haiti	39.8	45.2
Nicaragua	589.7	38.7
Zambia	191.3	174.4
Pakistan	49.5	35.3
Honduras	124.6	31.0
Indonesia	56.9	30.9
Morocco	71.0	32.1
Algeria	83.1	38.7
Brazil	121.6	37.9
Hungary	72.8	39.1
Argentina	33.1	34.7
China	(1996)17.1	9.8

Note: Countries of debt service of export of more than 30 per cent except China.

Source: World Bank (2001).

literacy rate, cereal yield, inflation rate and external current account. Institutional improvement is required even though structural adjustment programmes are in progress. Vietnam adopted the growth strategy of fostering three districts by establishing industrial zones in the 1990s (Kuchiki, 1995). The three districts are the northern district of Hanoi, Haiphong and Quang Ninh; the central district of Danang and Hue; and the southern district of Hochimin, Bien Hao and Vung Tau. Some of the industrial zones changed into export processing zones and succeeded in attracting foreign investors. However, inflows of FDI into Vietnam stagnated in the middle of the 1990s, partly because the government adopted a social policy in addition to the policy of inviting foreign investors. The social policy emphasized more equity than efficiency in order to mitigate the gap between the rich in the urban areas and the poor in the rural regions. The government encouraged the introduction of the Korean way of fostering big business groups (*chaebols*) as an industrial policy before the 1997 Asian currency crisis, but was forced to change the policy since it became clear that the Korean *chaebols* faced management related problems. Vietnam can focus on reforming state run enterprises and developing the stock markets in Step 5 of the growth strategy.

Table 6.9   External debt (percentage of GNI) and debt service (percentage of exports), selected countries

	External debt 1999	Debt service 1999
Algeria	198	37.8
Argentina	436	75.9
Bolivia	399	32.0
Burundi	1791	45.6
Chile	184	25.4
Columbia	223	42.8
Côte d'lvoire	238	26.2
Hungary	102	26.6
Indonesia	255	30.3
Lebanon	413	49.4
Macedonia	93	29.8
Mauritania	681	28.4
Mexico	105	25.1
Pakistan	342	28.2
Peru	358	32.6
Morocco	n.a.	24.3
Laos RPD	527	7.7
Vietnam	162	9.8
China	16	9.0

Note: Countries of debt service of export of more than 25 per cent except Morocco, Laos, Vietnam and China.

Source: World Bank (2001).

China

The government of China prioritized two major reforms of state owned enterprises and the banking sector. The tenth five-year plan of China in 2001 proposed to resolve the development of core industries in order to lead the national economy, narrow the gap of income inequalities and solve environment related problems. These reforms were the most urgent reforms in China in the 1990s. However, this country had a limited budget. Therefore, these reforms had to be prioritized in the following order.

According to Table 6.11, China is the best among the four nations analyzed in this section. It satisfies every criterion except for the fact that the Gini coefficient is nearly 50 per cent (see Table 6.11). It is well known that China is unequal in the incomes earned between the coastal area and the inland one. Thus, this country can focus on the growth strategy together with the reduction of poverty in the western region. The Chinese policies

Table 6.10   Gross domestic savings (Percentage of GDP)

	1993–95		1999
Algeria	27.9	**Algeria**	**31.8**
Angola	35.2	Belgium	25.1
Austria	25.7	China	40.1
Chile	28.0	**Congo, Rep.**	**29.7**
China	41.5	Czech	26.7
Gabon	43.7	Equatorial Guinea	57.8
Hong Kong	33.6	Finland	27.7
Indonesia	35.6	Gabon	34.8
Iran	30.3	Hong Kong	30.5
Ireland	27.6	Hungary	26.3
Japan	31.1	Ireland	37.0
Korea, Rep.	35.6	Japan	27.7
Malaysia	36.7	Korea, Rep.	33.6
Morocco	14.9	Macao	45.1
Netherlands	26.2	Malaysia	46.8
Oman	26.4	Netherlands	26.7
Papua New Guinea	33.8	Norway	30.3
Russian Federation	29.9	Russia	31.5
Saudi Arabia	28.5	Singapore	51.7
Slovak Rep.	26.9	Slovak Rep.	26.5
Switzerland	27.4	Taiwan	26.1
Thailand	36.2	Thailand	32.5
United Arab Emirates	32.6	Trinidad	26.6
		Turkmenistan	26.0
		Morocco	20.0
		Laos RPD	(1998)13.3
		Vietnam	23.2

Note: Countries of gross domestic savings of GDP of more than 25 per cent except Laos and Vietnam.

Source: World Bank (2001).

in the 2000s are consistent with our analysis. Table 6.11 shows that China is recommended to emphasize the policies of [F] institutional reforms, [H] environmental problems, and [G] growth strategy.

Applying the flowchart method to China on a province-wide, rather than nationwide, basis yields the results presented in Table 6.12. This has been done to show that there are poor provinces such as Xizang and Qinghai in inland China. Primary education is the first priority of these provinces. Table 6.12 also elucidates that some provinces have a surplus of electricity supply, while others have a deficit; this makes

Table 6.11   Data on Morocco, Laos, China and Vietnam (Unit %)

	Morocco	Laos	Vietnam	China	Step
Adult illiteracy (1999, %)	**52**	**52**	7	16	1
Cereal yield (2000, Kg)	**471**	3184	4048	4735	1
Current account balance of GDP (1999, %)	−3.3[1]	−6.1	−0.2	2.1	2
Inflation (1999, %)	5.8[2]	**126.3**	5.5	−1.4	2
Over all budget account of GDP (1999, %)	−1.4[3]	−4	−2.8	−2.9	2
External debt service ratio (1999, %)	32.1[4]	7.7	9.8	9	2
Energy production (1998, kwh)	508	n.a.	**283**	938[5]	5
Gross domestic saving (1999, %)	20	**13.3**[8]	**23.2**	40.1	5
Distribution of income	39.5[6]	30.4[7]	36.1	**40.3**	6

Notes: 1 - 1993–95; 2 - 1990–95; 3 - 1994; 4 - 1995; 5 - 1998; 6 - 1998–99; 7 - 1992; 8 - 1998.

Source: Tables 6.1–6.9.

Table 6.12 Data on provinces (2001)

	Number of schools (2000)	Number of graduates	Waste Water dis-charge	Waste Water discharge/GDP	Polluted air discharge (100 mcm)	Polluted air discharge/GDP	Industrial solid waste (10000 ton)	Industrial solid waste/GDP	Total market index (1999)	Market brokers and laws index	Adult illiteracy (%)	Grain production (kg/hectare)	Energy production (100mkwh)
Xizang	4	764	1006	7.3	15	0.11	17.1	0.10	–	–	47.25	5,044	–
Qinghai	7	2202	4661	15.5	607	2.02	336.7	1.12	2.00	2.37	25.44	3,361	111.90
Guizhou	23	13739	20598	19.0	3882	3.58	2271.8	2.09	3.86	1.59	19.85	4,236	335.19
Gansu	18	14255	23795	22.2	2800	2.61	1703.8	1.59	4.02	2.30	19.68	2,905	306.09
Ningxia	6	3154	10942	36.7	1445	4.84	478.7	1.60	2.69	3.24	15.72	4,044	151.81
Yunnan	24	18573	35117	16.9	2749	1.33	3187.4	1.54	3.39	2.16	15.44	3,923	320.75
Anhui	42	29830	63106	19.2	3945	1.20	2815.1	0.86	5.40	1.69	13.43	4,829	359.59
Neimenggu	18	12218	21844	14.1	4768	3.08	2375.6	1.54	6.45	2.74	11.59	3,854	280.89
Shandong	47	58355	110324	11.7	12179	1.29	5407.3	0.57	6.22	2.84	10.75	5,268	1,104.53
Sichuan	42	42672	116979	26.5	4779	1.08	4714.2	1.07	5.29	2.49	9.87	4,759	589.57
Shaanxi	39	36587	30903	16.8	2379	1.29	2625.0	1.42	4.48	3.40	9.82	3,095	321.54
Hainan	5	4021	7064	12.9	434	0.79	94.9	0.17	5.65	2.28	9.72	4,169	42.96
Fujian	28	24307	57617	13.5	2828	0.66	2190.5	0.51	7.28	2.70	9.68	5,089	439.19
Hubei	54	56566	106733	22.9	5674	1.22	2817.8	0.60	5.53	2.65	9.31	5,885	526.02
Chongqing	22	22187	84344	48.2	1908	1.09	1304.8	0.75	5.57	2.86	8.90	4,448	220.54
Hebei	51	43473	89600	15.5	9858	1.71	7027.9	1.22	6.70	1.96	8.59	4,025	867.55
Zhejiang	35	32477	136433	20.2	6509	0.96	1385.7	0.21	8.24	3.27	8.55	6,082	848.40
Henan	52	45709	109210	19.4	7436	1.32	3625.0	0.64	6.00	1.96	7.91	4,907	808.41
Jiangsu	69	75643	201923	21.2	9078	0.95	3038.2	0.32	7.04	2.78	7.88	6,302	1,078.44
Xinjiang	16	13774	15365	10.3	1944	1.31	718.2	0.48	2.90	3.74	7.72	5,771	197.92
Jiangxi	32	25903	41956	19.3	2220	1.02	4796.2	2.20	5.12	1.89	6.98	5,233	222.28
Tianjin	21	20112	17604	9.6	1749	0.95	469.8	0.26	6.58	5.33	6.47	4,820	247.94
Heilongjiang	35	35180	52644	14.8	4326	1.21	2693.7	0.76	3.97	3.53	6.33	4,512	456.86
Sanghai	37	40929	72446	14.6	5755	1.16	1354.7	0.27	6.59	6.25	6.21	7,320	592.98
Hunan	52	47426	112563	28.3	3569	0.90	2354.6	0.59	5.99	2.30	5.99	6,015	439.78
Liaoning	64	53353	109044	21.7	9432	1.87	7562.5	1.50	5.60	4.06	5.79	4,891	764.77
Jilin	34	30480	37386	18.4	3082	1.52	1604.4	0.79	4.51	3.22	5.74	5,125	295.08
Shanxi	24	20657	32406	18.2	6635	3.73	7694.5	4.32	4.57	3.72	5.68	2,806	557.58
Guangxi	30	21858	81571	36.6	4607	2.06	2108.1	0.94	5.28	2.32	5.30	4,656	331.92
Guangdong	52	51432	114055	10.7	8326	0.78	1694.3	0.16	8.33	4.93	5.17	5,360	1,458.42
Beijing	58	51931	23164	8.1	3227	1.13	1139.3	0.40	6.30	11.28	4.93	5,194	399.94

Sources: Chinese Education Yearbook, Chinese Environment Yearbook, Chinese Marketization Index Report 2000.

the reallocation of electricity between the provinces desirable. Using the province-wide analysis of the flowchart method, we can reconfirm that China does not face any problems with regard to grain and energy production.

Now, we focus our analysis on the policies of [F] institutional reforms, [H] environmental problems and [G] growth strategy. The indexes of marketization given in Table 6.12 represent the levels of [F] institutional reforms. The three indexes of wastewater, air pollution and industrial solid waste represent the levels of [H] environmental problems. The number of schools and graduates given as the indexes of higher education in Table 6.12 represent the levels of [G] growth strategy. Both the index of marketization and the index of market-intermediate organizations and laws given in Table 6.12 represent the levels of marketization in a country. Our criterion for the marketization of provinces is that each index is less than the average of each index. The three provinces of Guizhou, Gansu and Yunnan need to be marketized according to the criterion.

Table 6.12 illustrates the provinces of which each index of environment is more than the average of each index of the three environmental indexes mentioned above. The provinces that should give priority to the improvement of their environments are Gansu, Ningxia and Liaoning. We select the provinces that should emphasize higher education. According to Table 6.12, one index of higher education is the number of schools and the other is the number of graduates. The criterion for provinces to emphasize higher education is that we choose 11 provinces excluding both the ten provinces from the top and the ten from the bottom of each index of the two indexes. We chose the provinces by applying the criterion of the two indexes. Fujian, Shaanxi, Jiangxi, Heilongjiang, Shanghai, Jilin, Guangxi and Zhejiang need to emphasize higher education in their growth strategy. It is observed that the criterion of this chapter is not objective but arbitrary and remains to be examined by further studies.

Conclusion

There are three stages for economic growth, as shown in Figure 6.1. The first stage is to reduce poverty so that people can attain the social subsistence level. Poverty reduction policies at this stage prioritize primary education and enhancement of food production in order to attain the social subsistence level. The second stage is to build capacity in order to prepare for sustainable growth, which means growth with a sound environment. Capacity-building means to stabilize the macroeconomy, facilitate the infrastructure, regulate institutions and foster human resources. The third stage is to introduce growth strategies such as the industrial cluster policy, promotion of market competition, fostering of supporting industries, export promotion, introduction of FDI and so on.

It is observed that growth strategies will have negative effects of enlarging income disparities and worsening the environment. The policies of reducing the poor and improving the environment are inevitable for sustainable growth. In addition, it is desirable to change environmental policies according to the stages of development (Matsuoka and Kuchiki, 2003).

This chapter built a prototype model of how to prioritize policies in a certain order by using a flowchart. We divided the above three stages into the following six steps of deciding the priorities of policies: Step 1   attain the social subsistence level (primary education, health care, and food sufficiency); Step 2   attain macroeconomic stability; Step 3   liberalize the economy (trade, invest and finance) through structural adjustment programmes; Step 4   build capacity specific to growth strategies by facilitating a sufficient infrastructure (physical infrastructure and institutions); Step 5   adopt a growth strategy; and Step 6   narrow income inequalities.

We illustrated the effectiveness of our flowchart method on the case studies of Laos, Vietnam and China. Laos should not place too much emphasis on growth strategy at the expense of educational reform, attainment of macroeconomic stability and institutional capacity-building of Steps 1, 2, and 3, respectively. Vietnam can focus on the problems of reforming state run enterprises and developing the stock markets in Step 5 of the growth strategy. According to this chapter, this flowchart method should be applied to China on a province-wide rather than nationwide, basis. It is believed that the method presented should be compared with an alternative method since the criteria used are subjective.

References

Bai, X. and H. Imura (2000) 'A Comparative Study of Urban Environment in East Asia: Stage Model of Urban Environmental Evolution', *International Review for Environmental Strategies*, Institute for Global Environmental Strategies (IGES), 1 (1).

Burnside, C. and D. Dollar (2001) 'Aid, Growth, and Poverty Reduction', in *The World Bank: Structure and Policies*, ed. L.G. Christopher and D. Vines (Cambridge: Cambridge University Press).

CDF of Secretariat of World Bank (2001) *Meeting the Promise?* (Washington, DC: World Bank).

China Environmental Yearbook Edit Committee (2000), *China Environmental Yearbook* (Beijing: Environmental Yearbook Publishers).

Deininger, K. and L. Squire (1996) *New Ways of Looking at Old Issues: Inequality and Growth* (Washington, DC: World Bank).

ESCAP (1999) *Integrating Environmental Considerations into Economic Policy Making Processes* (Bangkok: United Nations Economic and Social Commission for Asia and the Pacific).

Furman, J. and J. Stiglitz (1998) 'Economic Crises: Evidence and Insights from East Asia', *Brookings Papers on Economic Activity* (Macroeconomics), 2 (Washington, DC: Brookings Institution).

Ito, T. (2001) 'Growth, Crisis, and the Future of Economic Recovery in East Asia', in *Rethinking the East Asian Miracle*, ed. J.E. Stiglitz and S. Yusuf (Washington, DC: World Bank).

Kawamura, M. (2002) 'Silk and the Japanese', www1.isc.senshu-u.ac.jp (accessed October 2002).

Kuchiki, A. (1994) 'Market Mechanism and the Role of Government in the Process of Economic Development', *Economic Analysis*, Economic Planning Agency, 137: 11–16.

Kuchiki, A. (1995) *Promotion of Industrialization and the Golden Triangle's Development, Dynamic Vietnam* (Tokyo: Institute of Developing Economies–JETRO).

Kuchiki, A. and M. Tsuji (eds) (2005) *Industrial Clusters in Asia* (London: Palgrave Macmillan).

Kuchiki, A. and K. Yamada (1997) 'Lessons from Japan: Industrial Policy Approach and the East Asian Trial', in *Economic and Social Development into the 21st Century*, ed. L. Emmerij (Washington, DC: Johns Hopkins University Press).

Kuznets, S. (1995) 'Economic Growth and Income Inequality', *American Economic Review*, 45: 1–28.

Lewis, W. Arthur (1955) *The Theory of Economic Growth* (London: Allen & Unwin).

MacIntyre, A. (2001) 'Rethinking the Politics of Agricultural Policy Making', in *The Evolving Roles of State, Private, and Local Actors in Asian Rural Development*, ed. A. Siamwalla (Hong Kong: Oxford University Press): 243–70.

Matsuoka, S. and A. Kuchiki (2003) *Social Capacity Development for Environmental Management in Asia* (Chiba: Institute of Developing Economies–JETRO).

Myint, H. (1965) *The Economics of Developing Countries* (New York: Praeger).

UNDP (2002) *Human Development Report 2002* (New York: UNDP).

World Bank (1993) *The Asian Miracle* (Washington, DC: World Bank).

World Bank (2001) *SIMA Query* (Washington, DC: World Bank).

7
Social Capacity for Infrastructure Management

Megumi Muto

Introduction

As in the case of environmental management, infrastructure management is a result of interplay between the government, civil society and the business sector. However, the case of infrastructure management may differ from that of environmental management in light of the following: (1) the roles of local and regional governments are more distinct; (2) the civil society tends to be divided into beneficiaries and the negatively affected; and (3) the business sector is less involved.

In this chapter, I first discuss social capacity as a repeated game amongst the government, the business sector and civil society, borrowing from game theory. Transactions that are repeated over a long period between the same actors may lead to a cooperative game resulting in improved management of the environment or infrastructure. In developing societies, where rules and laws are weak and difficult to enforce, cooperative game solutions are potentially effective ways to manage issues that involve externalities, such as the environment or infrastructure. Further, a review of discussions in the area of game theory suggests the importance of an institutional mechanism for coordination, retaining the same players over time, as well as accountability.

Addressing social capacity development (SCD) in the context of infrastructure management may be more challenging than doing so in the context of environmental management due to characteristics (1) and (2) above. The case studies in this chapter lead to a policy suggestion that under rapid urbanization, where a regional perspective needs to be introduced for the infrastructure, institutional coordination between local and regional levels as well as better accountability towards civil society are keys to success for infrastructure management.

In *Connecting East Asia*,[1] the Asian Development Bank, the World Bank, and the Japan Bank for International Cooperation jointly recognized the challenge of infrastructure coordination under the recent decentralization process in East Asia. In order to avoid fragmentation, it is essential to coordinate infrastructure both horizontally and vertically. The report underscores that 'where excessive fragmentation is a concern, efficiency gains may be achieved by clustering municipalities to form regional areas of service provision (albeit at the expenses of local accountability)'. The present chapter is an attempt to define a place for this infrastructure management concern in the discussion of SCD.

Theoretical framework

Physical infrastructure as public goods with externalities

Due to certain characteristics of a physical infrastructure, purely market mechanisms will fail to achieve efficient and sustainable provision of an infrastructure service. In the context of microeconomics, physical infrastructure such as roads and water supply are public goods or quasi-public goods with externalities. The market can achieve efficient resource allocations to 'private goods' for which private property rights are entitled. However, in the case of public goods, an indeterminate number of people can use them jointly (non-rivalness), and it is difficult to impose user fees (non-excludability). Thus, everybody tries to utilize such 'public goods' without sharing the cost (free-riders). Since anyone seeking profit would not produce these goods, the government should be the agent to provide them (Hayami and Godo, 2005).

Game theory and public goods – the possibility of a cooperative game solution

Game theory for public goods is an interesting analytical tool in considering the problems related to issues involving externalities. Analyses of public goods under game theory tend to focus on non-cooperative games illustrating free riding. In this chapter, I would like to review some discussions in the area of cooperative games, with regard to the provision of public goods. This leads to the analysis of necessary conditions enhancing environmental or infrastructure coordination.

I first cite from Gibbons (1992) the case of the 'infinitely repeated game with trigger strategy' to illustrate the possibility of a cooperative game. In infinitely repeated games, the main theme is that credible threats of promises about future behaviour can influence current behaviour. Considering the case that each of the two players cooperates until someone fails to cooperate (which is called 'trigger strategy'), there exists a solution where it is best for both players to cooperate at every stage.

Although being a case for two players, this infinitely repeated game model implies that cooperation is possible. The first feature of this model is that the game is repeated infinitely with the same players. However, in the 'real world' of infrastructure coordination, that is often not the case, as will be discussed in the following section. The existence of trigger strategy is the second feature of the model, which is to set a penalty in case the agreement is breached. An explicit penalty may be effective when contracts can be enforced. But when contracts cannot be enforced, which is likely in the case of a weak legal system, an alternative institutional mechanism will be a key for effective coordination.

Additional issues in game context: 'partners' or 'strangers'

For repeated games, the players must be the same ones over time. This may not be too difficult where the problem is limited within a municipality where the local government, businesses and civil society consist of the same members over time. However, in the case of infrastructure in particular, players may switch to another entity. First, there are cases where the government level in charge changes in response to the need to regionalize infrastructure management. Typically, when the pace of urbanization is rapid, infrastructure that has been under municipal government needs to be transferred to the next tier of government, such as regional or provincial government. Second, civil society is often divided between beneficiaries and those negatively affected. Beneficiaries of a physical infrastructure are typically widespread, while the negatively affected are the people at or around the construction site, and the divide varies by type of investment.

Keser and van Winden (2000) describe a related problem by comparing 'partners condition' and 'strangers condition' in experimental economics. 'Partners condition' is where the same small group of players play a repeated public good game, while 'strangers condition' is where players change group formations. This chapter suggests that players in the 'partners condition' contribute significantly more to the common public interest than do players in the 'strangers condition'. The suggestion from this experiment is that keeping the same players in the game is important to inducing a cooperative solution.

Additional issues in game context: 'complete' or 'incomplete'

In game theory, monitoring each other's behaviour is crucial for decision-making at the next stage. These are the cases of 'complete' or 'incomplete' information. A game has incomplete information if one player does not know another player's pay off, such as in an auction when one bidder does not know how much another bidder is willing to pay for the good being sold (Gibbons, 1992).

Matsushima (2002) argues that the possibility of one player monitoring the other's move is critical in the achieving of a cooperative solution. Repeated games are often compared to explicit contracts. When perfect monitoring among the players can be proved, a cooperative solution can be maintained by explicit contracts with legal enforcement. It suffices to include in the contract a provision for penalty upon any non-cooperative move. However, monitoring is costly in reality. Information can also be biased, and it is sometimes difficult to interpret the meaning of actions. When it is difficult for a third party to verify the obtained information, inducing cooperation by an explicit contract is ineffective. In such cases, implicit cooperation should be employed instead of explicit cooperation by contract.

This theory suggests that accountability – by way of transparency of information and providing credible explanations – has a positive role in inducing a cooperative solution. When it is difficult to obtain account-ability, it becomes all the more important to pursue a means of inducing implicit cooperation.

Features for successful social capacity for infrastructure management

Implications from game theory

The infinitely repeated game model among the same players implies that cooperation is possible. However, in the 'real world' of infrastructure coordination, players change.

The existence of a trigger strategy is the second feature of the infinitely repeated game model. But when contracts cannot be enforced, which is likely in the case of a weak legal system, an alternative institu-tional mechanism will be a key for infrastructure coordination. This is particularly important when it is difficult to obtain credible information.

In sum, the important features for successful social capacity implied by game theory are as follows:

- It is possible to achieve a cooperative solution, provided that an insti-tutional mechanism for coordination exists, instead of relying on an explicit penalty

- Retaining the same players is advantageous for achieving a cooperative solution
- Accountability helps in enabling a cooperative solution.

In the following sections, I will discuss infrastructure management in relation to the features above and the main agents; namely, the government, civil society and the business sector.

Government

Since the government is a permanent institution, it is an appropriate player of an infinitely repeated game.[2] However, the optimal unit of government – local or regional – may change over time. Regional bodies play a critical role in addressing concerns that transcend local boundaries and that are characterized by economies of scale and externalities such as infrastructure.

In the case of the water sector, when the local population is small, the water system can consist of, for example, ground-water pumps, a small treatment plant and distribution pipes. However, to keep pace with urbanization, it will be necessary to set up a large-scale water intake plant (such as a dam) and a larger treatment plant, as well as sewerage. The water may be sourced from the entire watershed, implying the importance of watershed management transcending jurisdictional borders. Therefore, the optimal unit of operation can change from local to regional over time and according to the nature of the infrastructure service.

Accountability tends to improve when the government unit is closer to civil society. However, when regional management is technically required for infrastructure, it becomes critical to strike a balance between accountability and infrastructure management.

Civil society

Who are the stakeholders of infrastructure in civil society? People who are directly affected, project beneficiaries and, perhaps, outside advocacy NGOs as well? The standpoints of these stakeholders can differ greatly. This divide within civil society is important when considering the case of infrastructure; however, in the case of environmental management, the potential divide may not be as sharp.

For example, local residents near a particular sanitary landfill may have strong opinions on facility expansion. On the other hand, the regional population may be longing for an effective sanitary landfill to serve the whole region. In addition, infrastructure debates can attract a variety of non-residents who may wish to join the discussion.

Business sector

In the case of environmental management, the business sector is clearly an active player. The firms are not only beneficiaries of better environment but also play an obvious role as polluters. In addition, a larger number of firms have come to consider involvement in environmental management activities to be part of their corporate social responsibility.

However, in the case of infrastructure management, the majority of firms are beneficiaries of the services. Firms who are potential providers of services do not constitute the majority of the local business sector. As a result, the role of the business sector may be weaker in the case of infrastructure issues than in the case of environmental ones.

Social capacity for infrastructure management – a hypothesis

Game theory has suggested the possibility of a cooperative solution, provided that an institutional mechanism exists. In addition, accountability and retaining the same players over time are important in order to enhance the cooperative solution. However, as discussed above, divides exist within each of the three agents concerned (see Figure 7.1).

Therefore, in order to assess social capacity for infrastructure management, it is important to consider the two levels within the government sector: local and regional. The existence of a functioning coordination

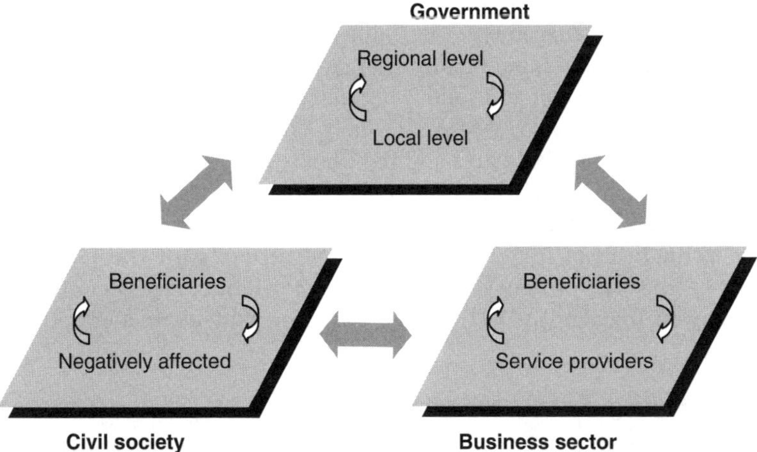

Figure 7.1   Social capacity for infrastructure

mechanism between the two levels is hypothesized as being a key aspect of social capacity for infrastructure management.

There may also be a need to divide civil society into two groups: the beneficiaries (often region-wide) and the negatively affected (often local); this need would arise when the divide between the two groups is significant. It is hypothesized that accountability on the part of the government may be institutional in minimizing the divide, thus constituting social capacity for infrastructure development.

Finally, as mentioned earlier, although we can identify both service beneficiaries and service providers, the role of the business sector may be weaker in the case of infrastructure issues than in the case of environmental ones.

In line with the above hypothesis, we now present case studies focusing on infrastructure management under rapid urbanization. These case studies examine the evolution of an institutional mechanism for coordination, including the issue of accountability. In sum, the cases in metropolitan Tokyo are those where a functioning institutional mechanism for coordination exists, although there are lessons for accountability. On the other hand, the cases in the Philippines can be summarized as having good potential for coordination and accountability but lacking a functioning institutional mechanism.

Cases in metropolitan Tokyo: the Tama area

The cases in metropolitan Tokyo are examples of how an institutional framework for coordination within the government sector is crucial in infrastructure management. The Tama area is of particular interest due to its experience of rapid urbanization and the variety in its approaches in dealing with water and solid waste management.

Metropolitan Tokyo is conventionally divided into three parts: the 23 wards area, the Tama area and the islands area (see Figure 7.2). Tama is geographically and historically distinct from the 23 wards. The former occupies the western part of metropolitan Tokyo, while the 23 wards are in the heart of Tokyo that occupies the eastern part. The Tama area had been a part of Kanagawa Prefecture, a neighbouring prefecture in southwest Tokyo until 1893, when it was incorporated into metropolitan Tokyo.

Water management in the Tama area

Water management in the Tama area has often been a major issue in metropolitan Tokyo. The Tama River originates in the deep forests of

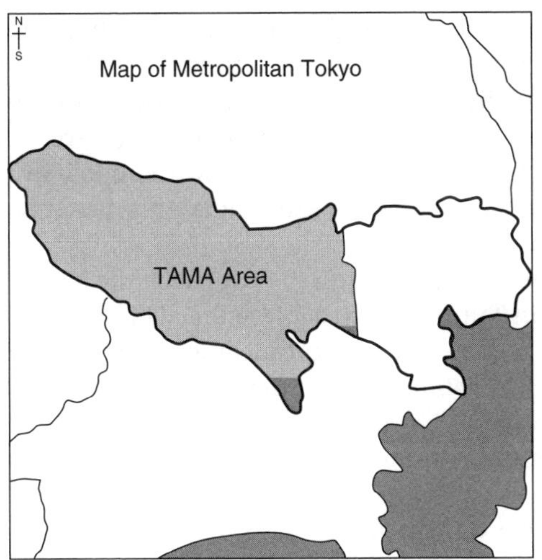

Figure 7.2   Tama area within metropolitan Tokyo

Tama and runs through the area, having long been a vital water source for people living in the 23 wards. The incorporation of the area into Tokyo was initiated by those heavily dependent on the water in the river, in order to preserve the natural water reservoir beneath the forest in Tama. The issue of water management in Tama was brought to the fore again by the unification of water management bodies in Tama.

For some considerable time, local governments in the Tama area had secured underground water, purified it and distributed it to every user within the jurisdiction on their own. The rapid urbanization in Tama, which accelerated in the 1950s, increased the demand for water, which led to the shortage of underground water and subsidence of the ground. In the meantime, the Water Law that came into effect in 1957 stipulates that local government should function as the water management body in its own jurisdiction. The local governments in Tama requested the metropolitan Tokyo government to distribute the surplus water to the Tama area when the area experienced a water supply shortage.[3] The higher price of the distributed water, which was due to the development of additional infrastructure, resulted in a widening price gap between the Tama area and the 23 wards.

The principle of the responsibility of a local government for water management began to crumble. The citizens and local politicians in Tama attempted to persuade the metropolitan Tokyo government to standardize water costs throughout Tokyo. Local governments had been concerned about the dismal prospect of water shortage under further urbanization. Some of the local governments in Tama filed a petition with the metropolitan Tokyo government authority to seek its assistance. Eventually, in 1971, the metropolitan Tokyo government formulated 'A Fundamental Plan of Unification by the Metropolitan Tokyo Government for Water Management in the Tama Area' based on the advice of an expert committee.

However, the strong objection to the plan by the labour union of local governments in Tama failed to achieve the planned unification of water management bodies. Political consideration produced a mode of water management that combined the unification of water management bodies as a basic approach, and at the same time allowed the consignment of a component of water management to local governments in Tama in order to secure the jobs of their employees.

The Tama area can be presented as a unique case of water management in Japan in that it is neither localized nor unified (see Figure 7.3). This is partly a reflection of Japan's Water Law. As stipulated in Act 6 of the Water Law, in principle, a local government is a body that assumes responsibility for water management within its jurisdiction. The preceding Act, on the other hand, makes a case for water management over the jurisdiction of local governments. The case study of Tama can be placed between the modes of water management expected by the two acts.

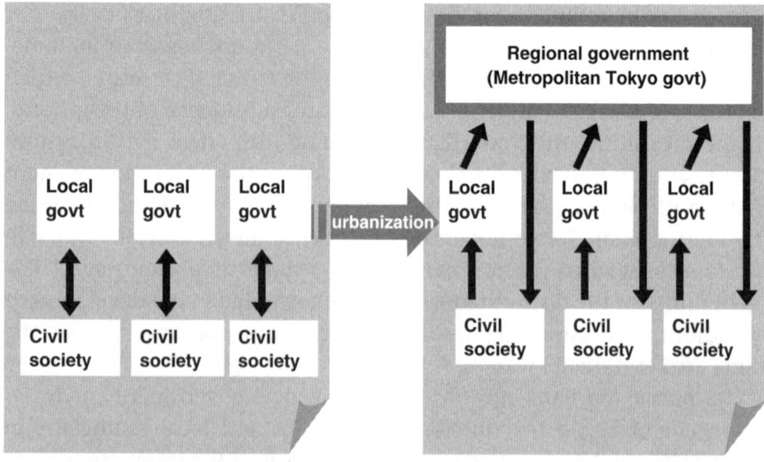

Figure 7.3   Water management in the Tama area

Widening the scope can be an optimum option for water management. The full use of scale merit enables more efficient resource allocation in the network type of infrastructure development and service provision. Moreover, the development of a large-scale infrastructure, often necessary in acquiring a new water source (for example, the construction of a dam and a treatment plant), is well beyond the financial capacity of a local government. On the other hand, the widening scope of water management can lead to a distant relationship between government and citizen. However, in the case of Tama, the local government remained as an accountability window for the citizens.

Water management in Tama is a case of a successful institutional framework of vertical coordination that keeps a channel for the citizen effectively to interact with the local government; this helped minimize conflict. At the same time, a water source was secured to meet the growing water demand of the citizens of Tama, through unification of the water management bodies with metropolitan Tokyo.

Solid Waste Management in the Tama area

Unlike the case of water management, solid waste management in the Tama area has been an issue of horizontal coordination among the local governments within Tama. The Solid Waste Management Law stipulates that a local government, in principle, is the body of waste management in their own jurisdiction. Each local government in the Tama area has historically self-managed the entire process of collecting, processing and sanitary landfilling. However, the rapid urbanization in Tama, which accelerated in the 1950s, increased the supply of solid waste, which led to the shortage of capacity for processing and landfilling. Thus, the principle of responsibility of a local government for waste management was challenged.

Article 284 of the Local Government Law, which became effective in 1947, allows for a special local government body for the execution of required services. In the case of waste management, more than two municipalities can jointly form a service cooperative and contract out the collection, processing and part of the sanitary landfilling services (see Figure 7.4). This service cooperative method became widespread in Japan because of its efficiency due to scale economies, and its safety due to the introduction of larger-scale and cutting-edge technology. However, it has been pointed out that the quality of residents' environmental awareness deteriorates as the services are not directly carried out by the municipality to which they belong (Taguchi, 1992).

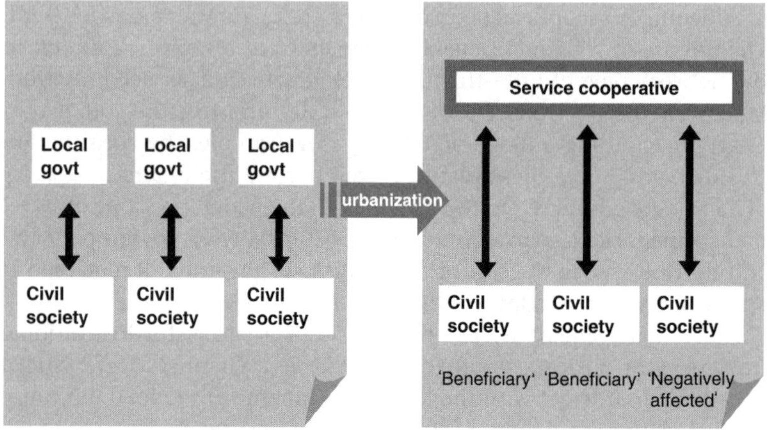

Figure 7.4   Solid waste management in the Tama area

The most controversial issue in Tama's solid waste management was information transparency. The two sanitary landfill sites in question are Yatozawa (used from 1984 to 1998) and Futatsuzuka (in use from 1998 to 2014). Residents questioned the safety of the Yatozawa sanitary landfill, which eventually led to a civil movement demanding that the safety of Yatozawa be secured and that the plan to construct Futatsuzuka be cancelled. The information dissemination process by the Tama service cooperative in response to this was not well-managed. The cooperative did not respond to the information disclosure request by the negatively affected citizens and proceeded with the Futatsuzuka plan. Frustrated by this act, a coalition of NGOs called the 'Hinode Forest Trust Movement' (membership of 2800) obtained the ownership of a part of the Futatsuzuka area as a 'trust common land'. This led to a forced eviction by the metropolitan government in 2000 in order to open the Futatsuzuka sanitary landfill.

As a result of this, some of the service cooperatives signed an agreement with the local government of Hinode to the effect that the metropolitan Tokyo government would pay Hinode a 'development fund' of 3.9 billion yen.

The service cooperatives' method for waste management had the potential to serve as a successful institutional mechanism for horizontal coordination among local governments with regional perspectives. However, the cooperatives' negative attitude toward information disclosure and the resulting negative reaction from civil society made this a case of accountability mismanagement worsening the divide within civil society.

Water management in Metro Cebu, the Philippines: from the perspective of the Tama area

Based on the two cases in the Tama area discussed above, it is suggested that a functioning coordination mechanism within the public sector is indeed important in infrastructure management, particularly when the perspective switches from local to regional. At the same time, accountability is important in order to minimize the divide within civil society. In this section, the case of water management in Metro Cebu will be reviewed from that perspective.

Government, civil society and business sectors in Metro Cebu

Metro Cebu is the second largest business center in the Philippines, hosting export processing zones and a great variety of industries, thus attracting population inflow from the neighbouring areas. According to the 1995 census, the area had a population of about 1.4 million. Metro Cebu consists of Cebu City, Talisay, Compostela, Liloan, Consolacion, Mandaue City and Lapu-Lapu City.

The Local Government Code of 1991 laid the foundation for local autonomy and the mechanisms for popular participation in local governance. Cebu City is known for its successful approach to a poverty alleviation programme in collaboration with civil society. This alliance was chosen as a best practice case of a GO–NGO partnership for poverty alleviation by the UNDP/UMP/Habitat II in 1994 (Etemadi, 2001).

Cebu's business leaders consider themselves as government partners in matters concerning development. The Cebu Chamber of Commerce Inc. (CCCI) is committed to contribute to the city's development agenda and to help address the pressing concerns of local government. It entered into a cooperation agreement with the Cebu City government in June 1998. Chamber members attend City Council meetings regularly, participate in consultations and public hearings, and continue to engage in dialogue with city officials (Etemadi, 2001).

There are over 30 development NGOs accredited by the Cebu City government. They focus on their services in the marginalized sectors in urban and rural areas. In addition to delivering basic services to communities, they emphasize community organizations in order to promote citizen participation and increase the capacity for self-help.

The case of Metro Cebu seems to have a high potential for collaboration between the government, the business sector and civil society. However, horizontal coordination within the government sector – that is, among the local governments – has not yielded tangible results.

Water management in Metro Cebu

As early as 1975, a metropolitan organization had been formed to develop and manage water supply and distribution in Metro Cebu. The Metro Cebu Water District (MCWD) is in charge of planning, designing, constructing, operating and maintaining water works within the seven areas of Metro Cebu. Funding is sourced through the Local Water Utilities Administration (LWUA) (Mercado, 1998). However, MCWD is not a coordination mechanism for local governments. It is a government-established public service provider covering several cities and municipalities.

Recent studies have shown that the water supply capacity of the MCWD is not sufficient for the growing urban population in the metropolis. Therefore, Metro Cebu is faced with the challenge of formulating a long-term plan aimed at solving the water supply problem. Within the framework of the Local Government Code, which enhanced the autonomy of local government units, it is not MCWD but the cities and municipalities that are expected to collaborate actively to formulate a common strategy. However, this has not yet happened.

A study by the Department of Environment and Natural Resources (DENR) and the Philippine Institute for Development Studies (PIDS) has called for a more integrated and holistic approach to the current fragmented and relatively weak institutional structure in order to handle water resource management in the metropolitan area and to address the inability to resolve conflicts related to inter-local government water transfers and inter-sectoral use of water (Mercado, 1998). The case of water management in Metro Cebu can be summarized as having good potential for coordination and accountability but lacking a functioning institutional mechanism led by the local government units. In response to the water crisis, many NGOs and business sectors are initiating small-scale watershed management or water supply projects. However, such activities lack the holistic approach necessary for effective water resource management (see Figure 7.5).

Solid waste management in Metro Manila, the Philippines: from the perspective of the Tama area

In this section, the case of solid waste management in Metro Manila will be reviewed. As learnt from the solid waste management case in Tama, not only coordination within the public sector but also accountability

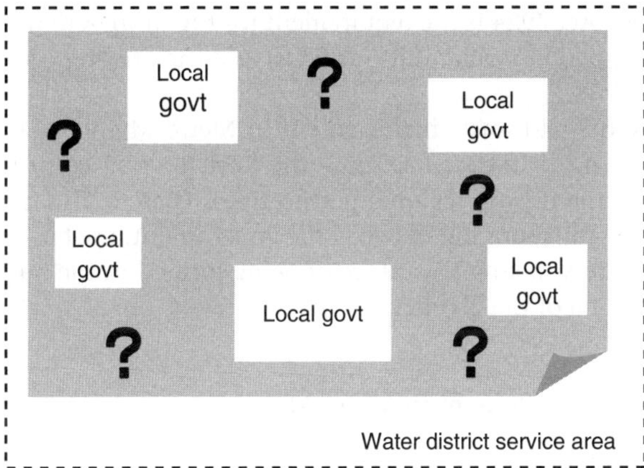

Figure 7.5   Water management in Metro Cebu

toward the civil society is important. This is particularly true in the case of sanitary landfills, which involve health and environmental concerns.

Government, civil society and business sectors in Metro Manila

Metro Manila consists of the cities of Manila, Caloocan, Las Pinas, Mandaluyong, Makati, Marikina, Muntinlupa, Paranaque, Pasay, Pasig and Quezon, as well as the municipalities of Malabon, Navotas, Pateros, San Juan, Taguig, and Valenzuela. In 1995, the total population of Metro Manila reached 9.5 million, accounting for 13.8 per cent of the total population of the Philippines. The Metro Manila Authority (MMA) was established in 1990 in response to the growing problems faced in Metro Manila. The MMA was considered as an interim body, pending the creation of a more permanent organization. It was also recognized as being weaker than its predecessor, the Metro Manila Commission (1975–89), in terms of both executive and revenue powers (Manasan and Mercado, 1999). The passage of the new Local Government Code (LGC) in 1991, which enhanced the autonomy of local government units, further weakened the MMA (Manasan and Mercado, 1999).

It was not until 1995 that the Republic Act 7924 was enacted; this Act defines Metro Manila as a 'special development and administrative region' subject to direct supervision of the President of the Philippines. The law also provided for the creation of the Metropolitan Manila Development Authority (MMDA). In general terms, the MMDA is in charge of planning, monitoring, coordinating, regulating and supervising the delivery of metro-wide services within Metro Manila. A distinct feature of the Republic Act 7924 is the requirement for the MMDA to work closely with NGOs, people's organizations (POs) and the private sector (Yu and Serrona, 2006).

The case of solid waste management in Metro Manila seems to be a case where an institutional set-up – the MMDA – exists for infrastructure management from regional perspectives. There is also tremendous potential for collaboration between the government, the business sector and civil society, taking advantage of being located in the business and political capital of the country.

Solid waste management in Metro Manila

Sanitation and solid waste management are basically the responsibility of the local government units as mandated by the Local Government Code. Under the current arrangement in Metro Manila, local government units are responsible for waste collection while the MMDA is responsible for final disposal. Collection coverage for 1997 among the local government units ranged from 40 per cent to 98 per cent. The bigger challenge lies in identification and maintenance of the final disposal sites. Previously, there were two existing landfills; Carmona and San Mateo. Both were closed by 2000 due to public opposition as a result of poor maintenance. According to a report by the Asian Development Bank, there were eight controlled dumpsites in the metropolis in 2003 due to be closed in 2006. Currently, the controlled dumpsites are under the responsibility of the local government units, some of which are contacted out to the private sector.

The MMDA, the local government units, and the various units of the DENR and DPWH take on various functions, jurisdictions and responsibilities regarding solid waste management. However, none of these bodies orchestrate all these activities to make them complementary rather than conflicting. An ADB report (Asian Development Bank, 2004) on solid waste management in Metro Manila highlighted the absence of a single agency assuming overall responsibility for all aspects of solid waste management from collection, transport and transfer to recycling and final disposal (Pardo, 1996). The inability of the MMDA and the local government units to coordinate their actions vertically has resulted in no tangible improvements in solid waste management. The case of waste management in Metro Manila can be summarized as having good potential for coordination and accountability but lacking a functioning institutional mechanism jointly led by the MMDA and the local government units (see Figure 7.6).

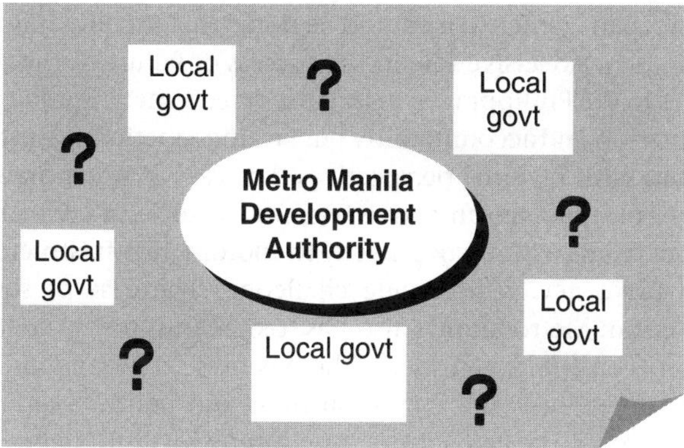

Figure 7.6   Solid waste management in Metro Manila

Conclusion

This chapter discussed social capacity as a repeated game among the government, civil society and the business sector, borrowing from game theory. In developing societies, where rules and laws are weak and difficult to enforce, cooperative game solutions are potentially effective ways to manage issues that involve externalities, such as the environment or infrastructure. Further, a review of discussions in the area of game theory suggests the importance of an institutional mechanism for coordination, retaining the same players over time, and accountability.

Addressing social capacity development in the context of infrastructure management may be more challenging than doing so in the context of environmental management. The roles of local and regional governments are more distinct; further, civil society tends to be divided into beneficiaries and the negatively affected.

The water and solid waste cases in metropolitan Tokyo (the Tama area) suggested that in the advent of rapid urbanization, a change in the government level is inevitable. If the optimal unit of government changes abruptly, it may negatively affect a cooperative relationship with the business sector and civil society. Therefore, the coordination within the government sector, including both regional and local bodies, is crucial. Metropolitan Tokyo seems to have managed the urbanization process by (1) introducing a system whereby the metropolitan level uses the local governments as agencies in tasks directly related to customers (water management), or (2) creating a cooperative consisting of local governments to pursue part of the task of a regional government (solid waste management). It should be noted that accountability to civil society can play a decisive role in the success or failure of the process.

The cases in the Philippines can be considered as having good potential for coordination and accountability but lacking functioning institutional mechanisms with regional perspectives. The case of water management in Metro Cebu involves an active business sector and civil society but local governments with weak capacity to coordinate infrastructure development with a new region-wide challenge. The case of solid waste management in Metro Manila also has tremendous potential for collaboration between the government, the business sector and civil society, taking advantage of being in the business and political capital of the country. However, the case lacks a functioning institutional mechanism to solve the problem of sanitary landfills serving the metropolis.

The case studies in this chapter lead to a policy suggestion that under rapid urbanization, where regional perspectives need to be introduced when developing an infrastructure, functioning institutional coordination between local and regional levels, as well as better accountability to civil society, are keys in developing social capacity for infrastructure management.

Notes

1 *Connecting East Asia* (2005).
2 Many governments change their behaviour after elections. This political aspect is beyond the scope of this chapter.
3 Historically, local governments in the Tama area did not have the right to acquire water from the Tama River, even though it runs through their jurisdiction.

References

Asian Development Bank (2004) *The Garbage Book* (Manila: Asian Development Bank).

Asian Development Bank, Japan Bank for International Cooperation, The World Bank (2005) *Connecting East Asia: A New Framework for Infrastructure* (Washington, DC: The World Bank).

Bureau of Waterworks, Tokyo Metropolitan Government (*Tokyo to suido kyoku*) (1994) *A Twenty-year History of Public Water Supply in the Tama Area* (*Tama chiku toei suido 20nen no ayumi*), Bureau of Waterworks, Tokyo Metropolitan Government (in Japanese).

Bureau of Waterworks, Tokyo Metropolitan Government (*Tokyo to suido kyoku*) (1999) *A Hundred-year History of the Tokyo Modern Water Supply* (*Tokyo kindai suido hyakunensi*), Bureau of Waterworks, Tokyo Metropolitan Government (in Japanese).

Bureau of Waterworks, Tokyo Metropolitan Government (*Tokyo to suido kyoku*) (2004) *Water Supply in Tokyo* (*Tokyo no suido*), Bureau of Waterworks, Tokyo Metropolitan Government (in Japanese).

Comprehensive Waste Disposal Corporative in Tama Area (*Tokyoto santama chiiki haikibutu koiki syobun kumiai*), Waste Disposal Corporative News (*Syobun kumiai nyu-su ekusupuresu*), http://www.tokyo-shobunkumiai.com/title.html (accessed on 13 October 2005) (in Japanese).

David, C., A. Inocencio, F. Largo, and Ed L. Walag (1998) 'Water in Metro Cebu: The Case for Policy and Institutional Reforms', *Journal of Philippine Development*, Philippine Institute for Development Studies, 46(25): 229–52.

Etemadi, F. (2001) 'Towards Inclusive Urban Governance in Cebu', *Urban Governance, Partnership and Poverty*, Working Paper 25, University of Birmingham.

Gibbons, R. (1992) *Game Theory for Applied Economists* (Princeton, NJ: Princeton University Press).

Hayami, Y. and Y. Godo (2005) *Development Economics: From the Poverty to the Wealth of Nations*, 3rd ed (Oxford: Oxford University Press).

Keser, C. and F. van Winden (2000) 'Conditional Cooperation and Voluntary Contributions of Public Goods', *Scandinavian Journal of Economics*, 102(1): 23–39.

Largo, F., A. Inocencio and C. David (1998) 'Understanding Household Water Demand for Metro Cebu', Discussion Paper Series 98–41, Philippine Institute for Development Studies.

Manasan, R. and R. Mercado (1999) 'Governance and Urban Development: Case Study of Metro Manila', Discussion Paper Series 99–03, Philippine Institute for Development Studies.

Matsushima, II. (2002) 'New Progress in Repeated Games: Implicit Collusion with Private Monitoring', in H. Imai and A. Okada (eds) *New Frontier of Game Theory* (*Ge-mu riron no sin tenkai*) (Tokyo: Keisoshobo Press), (in Japanese).

Mercado, R. (1998) 'Metropolitan Cebu: The Challenge of Definition and Management', Discussion Paper Series 98–15 (revised), Philippine Institute for Development Studies.

Pardo, E. (1996) 'Country Report of the Philippines', in A. Royston, C. Brockman and A. Williams (eds), *Urban Infrastructure Finance*, Asian Development Bank.

Special Research Committee for Reconstruction of Tokyo Water Project (*Tokyo to suido jigyo saiken chosa senmon iin*) (1968) *First Advice by the Special Research Committee for Reconstruction of Tokyo Water Project* (*Tokyo to suido jigyo saiken chosa senmon iin dai ichiji jogen*) (in Japanese).

Special Research Committee for Reconstruction of Tokyo Water Project (*Tokyo to suido jigyo saiken chosa senmon iin*) (1970), *Advice on Corrective Measures for the Disparity in the Water Project between 3 Tama Districts and 23 Special Wards in Tokyo* (*Tokyoto san tamachiku to 23tokubetsukubu tono suido jigyo niokeru kakusa zesei sochi ni kansuru jogen*) (in Japanese).

Taguchi, M. (1992) *Forefront of Garbage Problem* (*Gomi mondai saizensen*) (Tokyo: Sinnihon Publishing) (in Japanese).

Taguchi, M. (1994) *City Comparison of Contemporary Garbage Problem* (*Gendai gomi mondai no tosi hikaku*) (Tokyo: Sinnihon igaku syuppansya), (in Japanese).

Yorimoto, K. (1974) *Garbage War* (*Gomi senso*) (Tokyo: Nihon Keizai Shimbun Inc.) (in Japanese).

Yu, J. and K. Serrona (2006) 'Municipal Solid Waste Management in Metro Manila: Challenges and Options', *Macro Review*, 18 (1–2).

8
Capacity Assessment and Project Design in Environmental Management

Senro Imai, Taisuke Watanabe and Eiji Iwasaki

Summary

Designing a project on capacity development in environmental management

To design a project on capacity development in the field of environmental management, two capacities should be identified; namely, the current capacity and the required capacity (that is, the capacity to fulfil the requirement mandated by legislations and institutions and the needs of society). Typically, the following three steps are considered when designing a capacity development project.

Step 1: Understanding the general features of the current capacity of players and the factors influencing current capacity

The overall analysis of the features of the current capacity of players and the factors influencing it are discussed in the first step.

Step 2: Assessing the two types of capacity of legislations/institutions and players

1 Overall capacity assessment
The following two points are discussed:

- Why assess the capacity of legislations and related institutions for the enforcement of mandates?
- How to determine the observed and potential capacity of players.

2 Specific capacity assessment: capacity gaps between required capacity and current capacity

Capacity gaps are relative, since the level of required capacity varies according to the level of the targets set in a cooperation project. The issue concerning capacity gaps between current capacity and required capacity is discussed using the example of 'formulation of a pollution reduction plan'. Further, we discuss how to identify and set levels of required capacity and fill the capacity gaps.

Step 3: Setting the targets and framework for capacity development in project design

1 Setting the targets for capacity development
Features of the targets to be set in the cooperation project are discussed using the management levels, which have been divided into three classifications. This is followed by a discussion of the four points that should be considered when setting the targets.
2 Setting the framework of a capacity development project
Using two examples – a simple project and a comprehensive project – a project's progress in achieving the main targets is discussed. Further, a detailed discussion is presented on the provision of opportunities for developing capacity.

Design of a capacity development project: a case study on water quality management in the Philippines

The Philippines Clean Water Act (CWA) of 2004 mandated a comprehensive approach to address water quality problems by combining all the existing water quality management systems and introducing new policy instruments.

This case study describes the project design of the 'Capacity Development Project on Water Quality Management' in the Philippines. Further, this exercise is initiated based on the results of capacity assessment and the analysis of the CWA requirements. Elements related to the current capacity, required capacity, capacity gaps and aspects of project design are elucidated.

Background

JICA's support in the area of capacity development

The Japan International Cooperation Agency (JICA), which is responsible for technical cooperation in Japan, regards capacity development as the basic approach required to realize the Millennium Development Goals (MDGs). It defines 'capacity' as 'the ability to set and achieve goals and to identify and resolve development issues'. In other words, technical cooperation provided by JICA is a support that enhances the capacity of developing countries, enabling them to achieve the MDGs by themselves. In this context, a project for capacity development should be in accordance with the endogenous development plan of a country and integrated into the country's own programme.

In the last five years, JICA has conducted researches and studies in consultation with partner countries, donor agencies and academic institutions in order to ensure that technical cooperation projects for capacity development are carried out more effectively and efficiently. These researches and studies helped to identify some key elements such as the three-layered capacity (individual, organizational and institutional/societal) and the catalytic role of foreign experts in knowledge creation and ownership. In addition to these cross-sectoral activities, JICA has conducted thematic studies on capacity development and the nature of the sector in order to extend better suited supports to developing countries.

In this chapter, the term capacity development refers to the process in which individuals, organizations, institutions and societies develop 'abilities' – either individually or collectively – to respond to issues, perform functions, solve problems, and set and achieve objectives (JICA, 2004). The term capacity assessment refers to a structured and analytical process whereby the various dimensions of capacity are measured and evaluated within the broader environmental or systems context, as well as specific entities and individuals within the system (UNDP, 1997).

Moreover, in this chapter, the dimensions of capacity refer mainly to legal institutions and organizations.

Why capacity development in environmental management?

Since many developing countries have established legal frameworks for environmental protection, implementing/enforcing legal requirements, policy instruments and technical tools have recently assumed more importance. The capacity to implement and enforce legislation is especially critical; consequently, the workability/enforceability of legislation becomes even more important.

Bearing this in mind, the donor needs to examine the problems, identify the concerns of the partner and assess the capacity and willingness of the partner to solve these problems.

As a bilateral provider of technical cooperation, JICA is trying to elaborate project design in capacity development using capacity assessment. For example, in relation to pollution control, JICA introduced capacity assessment in the project design of a technical cooperation project that aimed at capacity development.

Designing a project for capacity development in environmental management through capacity assessment

To design a project for capacity development, two capacities should be identified – current capacity and required capacity. *Current capacity* represents the available capacity and forms the basis of the project design; *required capacity* refers to a capacity that is currently not available but needs to be developed through the cooperation project to attain certain, if not all, requirements stipulated by the legislations and related institutions for enforcement (hereafter referred to as 'leg/ins'). The following three steps have been developed, taking into account the experiences and considerations of JICA; in other words, these steps are considered appropriate and are partially employed when designing a capacity development project.

Step 1: Understanding the general features of the current capacity of the players and the factors influencing the current capacity

The main objective of capacity development is to develop the capacity of players (administrations, enterprises, citizens and academic/R&D institutes at individual as well as organizational levels) as well as that of the society, thus enabling them to deal with environmental issues.

Current capacity can be estimated by analyzing environmental activities. The environmental activities of the players are influenced by many factors such as socioeconomic conditions, features of the environmental issues that are being dealt with, and culture and tradition, as shown in Figure 8.1. The environmental activities are viewed as a reflection of these influencing factors. This viewpoint is also discussed in the concept and approaches developed based on 'Social Capacity for Environmental Management (SCEM)' (see Chapter 1), which deals with three social players – government, firms and citizens and their inter-relationships.

In order to examine capacity development in environmental management, we will closely study two dimensions of the factors – opportunity and constraint (see pp. 219–21). Moreover, we also attempt to identify how these factors actually affect the environmental activities and to what extent the capacity of players is demonstrated or restricted. It is evident that a player cannot act in isolation from other players; hence, the interrelationships among players will also be examined.

Through these series of analyses, it is possible to determine the factors that influence the activities of players, including opportunities and constraints that encourage and/or discourage players to display their capacity. Further, we can also identify the general features of the capacities of players (see Figure 8.1).

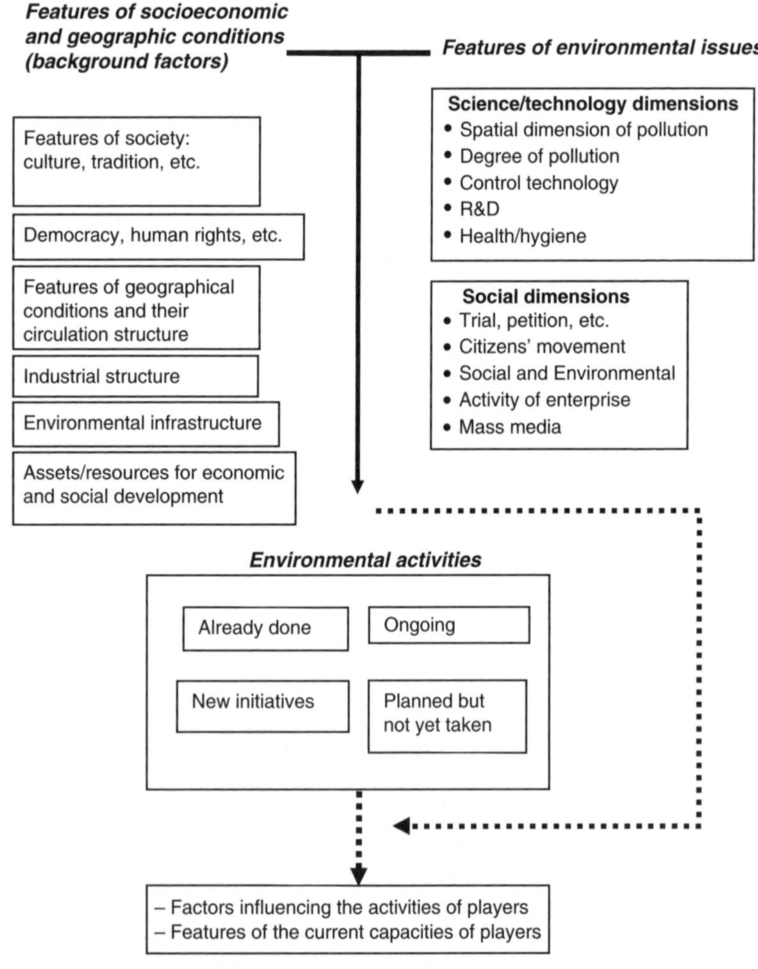

Figure 8.1   Environmental activities and the two influencing factors

Step 2: Assessing the two types of capacity of legislations/institutions and players

Overall capacity assessment

The overall analysis of the features of the current capacity of players and the factors influencing it, as discussed in step 1, is a preparatory step to conducting further analysis; namely, the two types of capacity assessments. Usually, the objective of the second step is clearly defined; that is, designing a technical cooperation project proposed by a host country. Although technical cooperation projects in the field of environmental management have different specific objectives, in general, the objectives are closely connected with the requirements stipulated by leg/ins. Many studies have discussed the weaknesses in the enforcement of the mandates stipulated by leg/ins. It is therefore necessary to pay attention to the cause of these weaknesses raised through the interactive relations between the leg/ins and players. Therefore, the focus of the exercise in this step is on leg/ins and players, and their interactive relations (see Figure 8.2 in page 211).

Why assess the capacity of legislations and related
institutions for enforcement of mandates?

The capacity of the leg/ins will be defined as their capacity to mobilize and develop the capacity of players. In case the leg/ins prove to be weak in terms of enforcement, legislations and systems or mechanisms are bottlenecked.

There is a strong tendency to attribute the weakness of leg/ins to the poor capacity of players. However, we have observed in many cases that the root cause of poor enforcement lies within the leg/ins. Causes of poor enforcement include factors hindering mobilization and demonstration of the capacities of players. For instance, the extremely stringent emission standards, which are unattainable given the existing technology and finance in enterprises, discourage the enterprises from abiding by the standards. Further, enterprises demonstrating good environmental efforts are often under-valued by administration, and this deters them from sustaining their efforts. The inability of administration to evaluate the situation accurately hinders appropriate provision of technical and financial support to the enterprises entitled to receive such supports.

Moreover, poor disclosure of the environmental performance of enterprises to the public fails to improve the behaviour of enterprises. Citizens are neither able to choose the products of enterprises making concientious environmental efforts, nor are they able to take action against enterprises who do not employ environmental efforts, thus allowing enterprises with no environmental policy to compete in the market and condoning the lack of environmental action. These are examples of how leg/ins are unsuccessful in providing opportunities to mobilize the capacity of players.

How to assess the observed and potential capacity of players

Typically, the capacities of players are assessed on the basis of observation and analysis of the players' performances in achieving the mandates stipulated by leg/ins. Such an assessment is critical for administrations and enterprises, which are core players in the enforcement of legislations. Important aspects of assessment with regard to other players, such as citizens and academic/R&D institutes, include the opportunities provided by the leg/ins to participate in environmental management and the utilization of the supporting measures under the current leg/ins. It has often been observed that citizens, NGOs and CBOs are actively taking initiatives to compensate or overcome the weaknesses of the administrations. This kind of creative display of capacities requires adequate attention, particularly for the future improvement of the current institutions.

As mentioned before, the current leg/ins in developing countries are ineffective in providing such opportunities. Therefore, the observed capacities often do not represent the actual potential for development of the players' capacities. Moreover, we believe that the performance of players can improve greatly if they are provided with better opportunities. In the following section, we examine the gap between the current and required capacities and discuss how to fill this gap. For instance, if the potential capacity is high, then the required capacity can be set at a considerably high level, despite the low level of the current capacity. Enhancing the current capacity to the required capacity level implies realizing the potential capacity by providing the necessary opportunities. Therefore, efforts to identify the potential for development in capacity are undeniably important in designing and implementing a cooperation project.

Specific capacity assessment: capacity gaps between required capacity and current capacity

This section discusses the identification of gaps between the required and current capacities. When designing a cooperation project, the consideration and setting of the requirement level is of crucial importance. This is because the gaps can be identified only after the requirement levels are set by comparing the required levels of capacity with the current capacity.

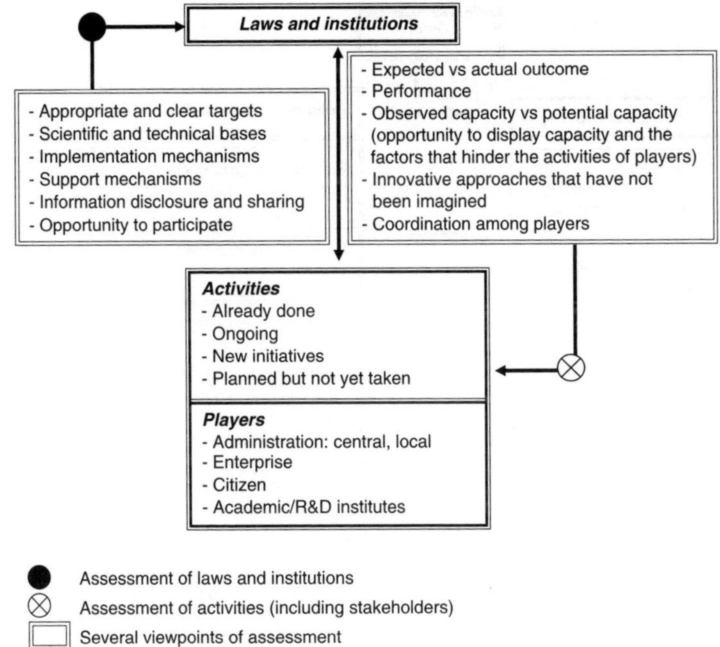

Figure 8.2   Two types of assessment

What are capacity gaps?

As pointed out earlier in this chapter, the targets set in a technical cooperation project are closely related to the requirements stipulated by the legislations and institutions for enforcement. Figure 8.3 illustrates the typical gaps between the current capacity often observed in developing countries and the requirements by leg/ins.

Figure 8.3 shows that there are many requirements that need to be satisfied in order to achieve the final objectives of the legislation and institutions; namely, a high compliance rate with standards and a good quality of environment. These requirements range from simple and technical requirements to sophisticated requirements (for example, formulation of a comrehensive pollution reduction plan or environmental performance rating for enterprises). Further, there are close interrelationships among the requirements; for instance, a sophisticated requirement such as the formulation of a pollution reduction plan needs some basic

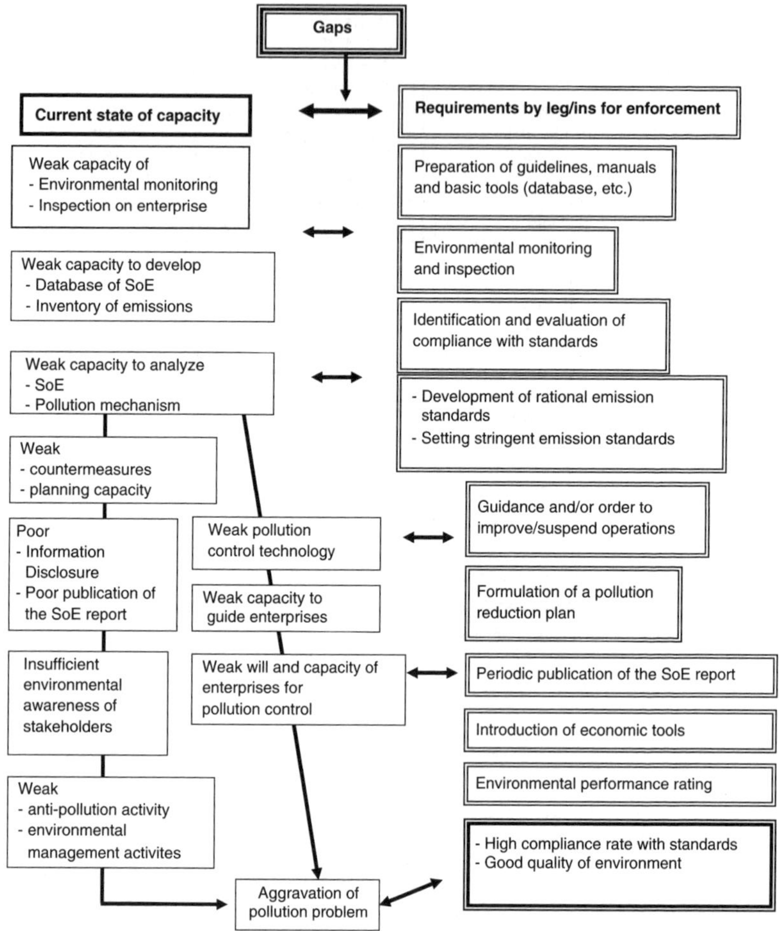

Figure 8.3 Capacity gaps between current capacity and requirements by legislations/institutions (a case study of 'pollution control' and 'management of pollution' (refer to Table 8.1))

requirements such as a monitoring database, an emission inventory and simulation models.

It is necessary to pay attention to the fact that the requirement level is not determined only by the related mandate of legislation. For instance, typically, the mandate related to the formulation of a pollution reduction plan merely states the need of the formulation and does not provide technical details. Detailed descriptions are usually provided by the

Implementing Rules and Regulations (IRRs). Typically, the IRRs specifies the level of the plan and, consequently, provides the basis for setting the level of the capacity to be developed in a cooperation project for the formulation of the plan. The same applies to other requirements, including the level of environmental monitoring and inspection, the State of the Environment (SoE) report and the introduction of economic tools, as shown in Figure 8.3.

These ranges within each requirement reveal that the capacity gap cannot be defined in isolation from the specific level of requirement to be selected or considered appropriate for the cooperation project. Thus, capacity gaps are relative.

Further, requirement levels are not even and balanced in relation to each other. For instance, in some developing countries, the ambient air monitoring is up to standard but the inspection of the enterprises is poor. In another instance, the ambient monitoring is well conducted, but the disclosure of the monitoring results is poor, and so forth. This sort of imbalance is widely observed, and it poses a difficulty in setting the requirement levels in a cooperation project.

How to set the requirement level and fill the capacity gap

Since the capacity gap is relative, it becomes clear only when we select the specific requirement level. Figure 8.4 illustrates how the capacity gap can be defined for the formulation of a pollution reduction plan. As described earlier, the requirements for the formulation of a pollution reduction plan range from basic to advanced. These requirements cannot be randomly selected; their selection is based on whether the necessary conditions are available. For example, if the current conditions are at the basic level, then a medium level requirement is considered possible and appropriate, provided proper technical assistance to raise the current capacity from a basic level to a medium level is extended. In this case, the capacity gap is between basic and medium level capacities. Setting an advanced requirement level when the current capacity level is basic will result in the widening of the capacity gap.

Given the limitations of time, input of experts and finance, setting a challenging level of required capacity poses some difficulties to both the implementing agency and the donor. However, often the host country insists on setting a challenging level of required capacity, while the donor prefers a practical and feasible level. If a challenging level is found to be necessary, then setting two to three steps to reach the challenging capacity level is worth considering, provided the host country is committed, devoted and well-equipped (staff, finance and so on).

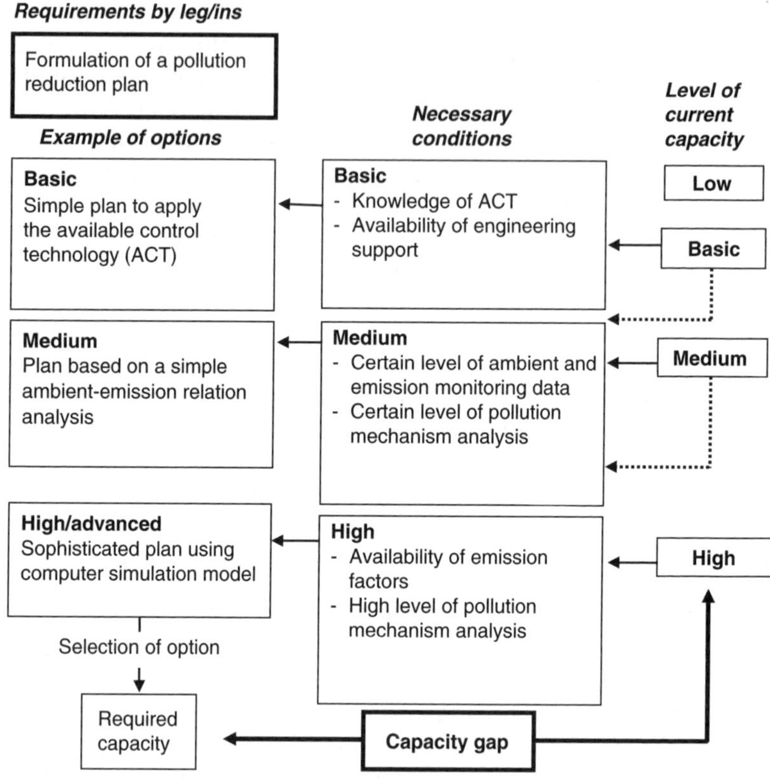

Figure 8.4 Relations among requirements, options, necessary conditions and levels of capacity: a case study of 'formulation of a pollution reduction plan'

Once the requirement level for a major or core capacity is set – for instance, formulation of a pollution reduction plan – then the level of related requirements – that is, conditions to formulate the plan – will be set accordingly. The same applies to other core capacities, such as the capacity to prepare the SoE report.

Setting the targets and framework for capacity development in project design

Setting the targets for capacity development

Targets for capacity development in a cooperation project vary according to the main purpose of the cooperation project. Table 8.1 attempts to

Table 8.1 Classification of the levels of management and their parameters

	Pollution control	Management of pollution	Environmental management
Basic concept	• Specific pollution source (point) as main target • Penalty • Regulation/inspection	• Spatial approach • Special measures for hot spot area/urban areas • Encouragement of voluntary measures • Participation of stakeholders	• Circulation of materials and pollutants in watershed/air shed • Management of hinterland environment • Carrying capacity
Space	• Specific pollution source	• Meso scale • Urban area	• Watershed, air shed • Nation-/region-wide
Object of counter-measures	Large-scale enterprise	• Large/medium scale enterprise in hot spot/urban area • Various type of pollution at urban level including mobile source	• Large-scale + SMEs + non-point source • Various type of activities affecting ecosystem
Stakeholders to be controlled	Large-scale enterprise	Large-medium-scale enterprise, automobile	Large-/medium-/small-scale enterprises, household, communities and people engaged in agriculture, fishery, forestry, etc.
Legal instruments	• Basic law for pollution control • Regulation laws • Licence	• Delegation of authority to local governments • Ordinance by local government • Law for total mass regulation in hot spot area	• Basic law for ecosystem or environmental management • Special law/ordinance for watershed/air shed environmental management

Continued

Table 8.1 Continued

	Pollution control	Management of pollution	Environmental management
Technical instrument or management tool	• On-site inspection system • Pollution control officer system • Database, emission inventory (simple)	• Pollution control plan for hot spot/urban area • Env. model city programme • Industrial park • Pollution prediction model (medium level)	• Wide area pollution load calculation model • Wide area pollution prediction model (advanced) • Land use plan based on the satellite image/GIS • Comprehensive database
Economic instrument or management tool	• Fines • Pollution control agreement	• Development of urban env. infrastructure • Pollution charge system • Env. performance rating system for enterprise	• Economic valuation system for ecosystem management • 3R based industrial restructuring plan
Information	• Basic database • Compliance rate with emission stds • Simple SOE report	• Load-ambient analysis related data and database • Statistics of waste, pollutants in areas • Env. performance related information • SOE report with analysis	• GIS information • Integration of socio/economic and env. information • Information on circulation of materials/pollutants • Comprehensive SOE report

classify the various purposes of projects on the basis of their management levels.

This table does not intend to indicate the course of transition from a simple management level (for example, pollution control) to a comprehensive level (for example, environmental management); rather, it intends to provide insight into the issues related to target setting.

Table 8.1 shows that there are several factors involved in setting targets for capacity development in environmental management.

(i)   *Targets to be set vary according to the main purpose of a cooperation project or the main need of a host country.* For instance, in the case of pollution control, the target will be simple, whereas for environmental management, the targets will be comprehensive.

(ii)   *Targets for capacity development are inherently comprehensive since capacity development cannot be achieved based on a single capacity area.* Even in a cooperation project with a simple target, several capacity areas are required. Table 8.1 indicates the range of capacity areas required for a simple cooperation project aiming at developing the basic pollution control capacity. Although the cooperation project has a simple target, several capacity areas are necessary, including the pollution control technology area, basic management tool development area (database, emission inventory) and the planning area (pollution control plan for a few large-scale enterprises). In case of a cooperation project with a comprehensive target – for instance, environmental management in water basin management – capacity development requires many capacity areas and the level of the capacities is advanced, not basic. As a result, various types of environmental management tools, including economic instruments, as well as coordination mechanisms to ensure the involvement of a wide variety of players need to be developed.

(iii)   *When setting targets, it should be considered that the level of related capacities (current capacities) is not even.* For instance, in some cases, ambient environmental monitoring is well performed, but inspection activities are poor, and vice versa. In other cases, both the ambient environmental monitoring and inspection activities are good, but the interpretation and disclosure of the monitoring results are ineffective, and vice versa. Therefore, when setting the targets, this kind of imbalance of related capacity levels should be taken into account.

(iv)   *Given this imbalance in the level and multi-strata relation of related capacity areas, it is not easy to identify and set proper targets.* Therefore, it is appropriate to set main targets that reflect the core objective of the cooperation project and attempt to set subtargets that support the main target.

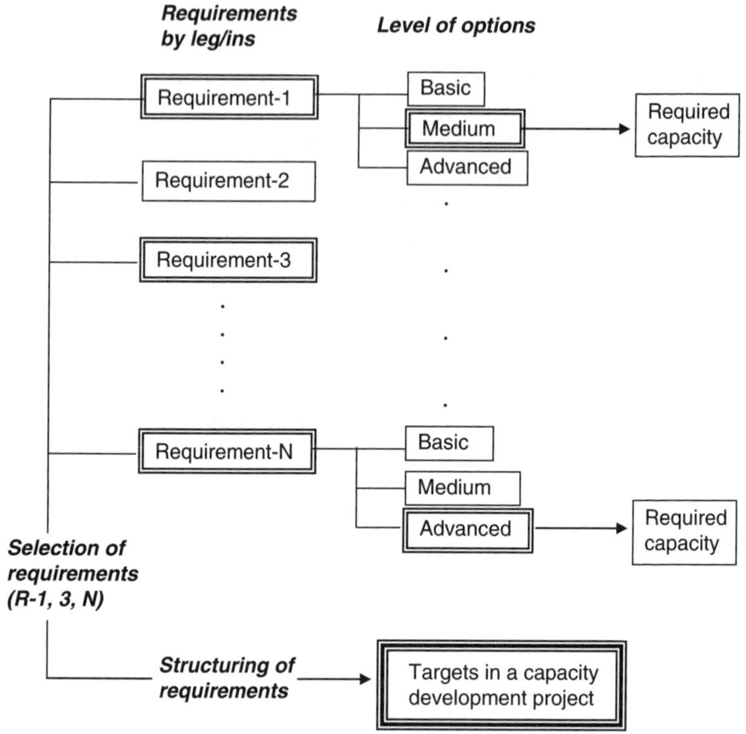

Figure 8.5   Interrelation of the targets in a capacity development project with requirements by legislation/institutions and required capacity

Figure 8.5 illustrates the interrelation of the targets in capacity development, the requirements and the required capacity. Each target consists of several requirements. For instance, in the case of a simple project, there will be one target that consists of several requirements, and in the case of a comprehensive project, there will be several targets, each containing several closely interrelated requirements. The project on the Philippines Clean Water Act (CWA) is a comprehensive project that has four main outputs, which will be regarded as the targets for capacity development. Under these four outputs, there are a total of 39 activities. The capacity to undertake each activity will be regarded as required capacity. The project of Philippines CWA is introduced in detail in the next section and will reveal the actual process of setting targets in a cooperation project in collaboration with the Philippines counterpart agency.

Setting the framework of capacity development in project design

Framework of a cooperation project is diverse and needs flexibility

The framework of technical cooperation projects is highly dependent on the set targets; thus, it is diverse. As expected, the framework of a project will be simple if the target is simple – for instance, the framework of capacity development in environmental monitoring. However, the framework will be complex if the targets are comprehensive, as is the case with capacity development in the Philippines, discussed in the following section.

The course that a project adopts to reach the main target is thoroughly examined in the project design. In this sense, the framework of the project is fairly fixed; however, at the same time, the framework should allow for some flexibility. For instance, even in the case of a simple project with a simple target, the framework might contain elements that increase its effectiveness. For example, in a project for strengthening the capacity in environmental monitoring, the elements that enhance the capacity of interpreting the monitored data further enhance the capacity to prepare a well-designed SoE report, thus ensuring the enhancement of the awareness of players toward environmental issues. Enhanced awareness and the disclosure of well-interpreted data will result in the recognition of the importance of environmental monitoring and strengthen the desire to improve the quality of monitoring. This, in turn, will exert pressure on the administration for improving monitoring activities. In a complex project with comprehensive targets, the key focus areas are the stratum structure and the interrelationship between several requirements. The timeframe for such a project will be longer and, in the course of implementation, new relevant requirements may possibly emerge. If a new requirement is found to be necessary, it is taken into account while framing the next phase of the project. This contributes to enhancing the effectiveness of cooperation.

A good cooperation project is one that makes progress and, during the course of progress, reveals higher level requirements. In this regard, the framework of a project is flexible. If the new requirement is found to be necessary, it is incorporated in the cooperation project. However, if it is found that the new requirement can be dealt with by the host country based on the achievements of the cooperation project, the cooperation is terminated. As explained by Matsuoka, the host country progresses from the 'system creation stage' to the 'self-management stage' (Matsuoka,

2006). Thus, the framework of a cooperation project needs to be flexible and capable of responding to these changes and progresses.

What kind of opportunities can be incorporated into the framework of a cooperation project

Capacity development cannot be achieved through mere paperwork. Donors are required to assist the efforts of players in the host countries. Donors can provide the players with opportunities to (i) plan, apply and verify without assistance; and (ii) realize the potential capacity, which is discussed on page 210.

In addition, the opportunity to eliminate constraints is vital. It is very important, both conceptually as well as practically, to incorporate these opportunities into the project. The following are some examples we usually consider.

(i) Utilization of a coordination mechanism The establishment of a coordination mechanism – for instance, between the line ministries – is usually an important objective of the cooperation project. However, the establishment of a coordination committee is only an initial step in terms of capacity development. Specific suggestions or plans to ensure that the coordinated actions are taken by the line ministries should be presented to the coordination committee. The capacity to prepare such suggestions or plans is needed, and the opportunity to develop such a capacity should be incorporated into the project.

(ii) Field testing of management tools The development of management tools – such as a database, emission inventory, manuals and guidelines for ensuring the implementation of economic instruments, or command and control – is required. However, it is very important to ensure the feasibility and validity of these tools in terms of science, technology and finance by verifying them. Thus, field testing of these tools is critical. Opportunities to prepare these tools at a central administration level, to ensure participation of local administrations, to conduct field tests, to verify and improve these tools based on the field tests, and so on should be incorporated into the project.

(iii) Participation-based activity The participation of players is essential in terms of ownership, effectiveness, and coordination. For instance, in the course of preparing a pollution reduction plan, participation of the following players is critical: enterprises as a target group of regulation, academic/R&D institutes as a supporting group to ensure scientific and technical validity, and citizens as a group of judges to determine

the appropriateness of the expected improvement. Opportunities that ensure such participation should be incorporated into the project.

These opportunities should be incorporated into the project at the appropriate time during the course of achieving the targets.

Design of a capacity development project: a case study on water quality management in the Philippines

Introduction

The capacity development project on water quality management in the Philippines, 'Capacity Development Project on Water Quality Management' by JICA and the Environmental Management Bureau (EMB), is explained as an example for designing capacity development projects, focusing on the implementation of a new legislation. This section focuses on the legal and organizational factors of capacity and does not cover the external factors of capacity (political, economic, social and cultural) outside the targeted institution.

Cognizant of the challenges faced in the enactment of the Clean Water Act (CWA), the EMB requested the Japanese government for technical assistance in a capacity development project on water quality management. When designing the framework of the project, the focus of the project was capacity development of the EMB and not of the other agencies that also had mandates under the CWA. This is because the capacity of the EMB and its regional offices is quite limited, while the CWA provides EMB and its regional offices with a number of mandates. Moreover, the EMB is the core agency in charge of implementing the CWA, and it plays the role of coordinator among these agencies. Capacity assessment was conducted based on an analysis of the water quality management activities; namely, implementation of legal requirements and management of the organization and resources of the EMB.

Water pollution

The Philippines is an island country, and its major cities are located at the mouth of a river. Although the cities are densely inhabited, the sewage system covers only a limited area. For instance, even in Metro Manila, the sewage system covers only 8 per cent of the population (World Bank, 2003). As slums are formed along the river, solid waste and waste water directly goes into the river. Industrial effluents are also a cause of water pollution. In particular, compliance with effluent standards is poor in small and medium-sized industries. Thus, water quality has progressively deteriorated and the dissolved oxygen levels are zero in some areas.

The environmental and public health dimensions of the water quality situation in the Philippines are as follows (World Bank, 2003):

- 36 per cent of the river sampling points have been classified as public water supply sources
- About 60 per cent of the country's population lives along coastal areas and contribute to the discharge of untreated domestic and industrial waste water from inland
- Preliminary data indicate that up to 48 per cent of ground water, intended for drinking water, is contaminated with total Coliform and would need treatment
- 31 per cent of illnesses over a five-year period were water-related deseases.

Clean Water Act

Water quality management (WQM) started by Presidential Decree 984 in 1976, and the National Environmental Protection Council and National Pollution Control Commission were established.

The Department of Environment and Natural Resources (DENR) and EMB were established in 1987 and issued DENR Administrative Order 34 (series of 1990) amending water classification/water quality criteria of 1978 and DENR Administrative Order 35 (series of 1990) amending effluent regulations of 1982.

However, the enforcement capacity of the EMB was limited and it became obvious that the command and control approach was insufficient. As a result, a new approach, a market-based instrument and self-regulation, was introduced in 2002 and 2003 by administrative order.

After four years of discussion, the Philippines Congress legislated the Clean Water Act (CWA) in 2004. The CWA greatly transformed the existing WQM framework.

First, the CWA provided mandates to the EMB as well as to related agencies, such as the Department of Health, Department of Public Works and Highways, and Local Government Units. Further, the CWA mandated the formulation of an integrated WQM framework, which is a policy guideline that includes water quality goals, periods of compliance, water pollution strategies and techniques, water quality information and human resource development. This integrated WQM framework was mandated to ensure the involvement of other agencies and stakeholders in the planning and implementing of water quality improvement actions.

Second, the CWA prescribed an area-wide approach. Thus, water quality management areas (WQMAs) were designated and the presiding governing board was responsible for coordinating policies and action plans. Further, non-attainment areas were designated, and related agencies were in charge of taking measures to upgrade water quality.

Third, the CWA introduced market-based instruments, such as the waste water charge system and the Water Quality Management Fund. In addition, effluent quotas were allocated in the discharge permits.

The following are other salient features of the CWA:

- Industry-specific regulation
- Review and revision of water quality guidelines and effluent standards
- Voluntary pollution prevention programmes (self-monitoring)
- Sewerage and septage management
- National Water Quality Status Report
- Fines and administrative action.

Current capacity

Outline of the EMB

The Environmental Management Bureau (EMB) of the Department of Environment and Natural Resources (DENR) sets and enforces standards, criteria, guidelines, inspections and monitoring for all aspects of water quality management (excluding drinking water). Further, the EMB is in charge of controlling pollution, including air pollution, and managing waste and toxic substances.

The EMB's functions and resources increased after the enactment of the Clean Air Act in 2001. As a result of the Clean Air Act, the EMB transformed from a policy bureau to a line bureau, with offices at regional level. As a line bureau, the EMB is involved in the operational aspect of environmental regulations. In 2003, the EMB had a staff size of 585 including supporting staff, and in 2001, its budget was 275 million pesos (5.39 million US$). The organizational structure of the EMB comprises a central office (CO) and sixteen regional offices (ROs). WQM functions are specifically carried out by the Water Quality Management Section (WQMS) under the Environmental Quality Division, in both the CO and ROs. The ROs are in charge of enforcement of legal regulation.

Implementation of existing laws

To guard against the environmental impacts of water pollution, the Philippines has many water-related laws and thirty years of experience; however, its enforcement is weak. The following are typical issues in enforcement capacity (World Bank, 2003):

- Inadequate government resources (that is, budget, manpower and facilities). For example, the EMB has not received any additional budget and continues to receive a small percentage of the DENR's annual budget despite the passing of additional laws it is mandated to enforce.
- Incomplete database. The EMB only has 25,000 (3 per cent) of the 826,783 firms registered in the country entered into its database. Of the 25,000 firms, only 14,111 (46 per cent) were inspected in 2001.
- Inadequate guidelines. Formal guidelines and plans to enforce laws are inadequate and sometimes absent.
- Lack of coordination among various agencies.
- Limited access to information due to the lack of a comprehensive, long-term environmental quality monitoring programmes.

Assessment of the current capacity

For the preparation of the project, JICA and the EMB conducted a baseline survey on the ROs' capacity (see Table 8.2), a questionnaire survey on the enforcement capacity, a field visit to the ROs, focus group discussions and key informant interviews. These tools were also used to study the required capacity.

The surveys revealed information pertaining to the current capacity and the critical weaknesses of the EMB. Among them, the important ones are as follows:

WQM policy   The capacity of the EMB in research and policy formulation is confined to the development of regulatory standards and the promulgation of rules and regulations. Before the enactment of the CWA, the objective of the WQM policy was to ensure that the point sources complied with the effluent standard. In addition, the EMB outsourced the drafting of policy documents, such as administrative orders, to consultants because of manpower shortage (Ohta and Kojima, 2003).

Water quality monitoring and water body classification   Water quality monitoring involves the monitoring of major water bodies at least once a year. A total of 118 water bodies are being monitored by the ROs and the CO. However, this accounts for approximately 26 per cent of the 457 classified water bodies nationwide. There are 184 principal rivers, approximately 44 per cent of total number of principal rivers, that still need to be classified (JICA and EMB, 2005). Presently, the activities of the ROs in relation to classified water bodies are limited to enforcement of the effluent standard and regular periodic monitoring of water quality. Thus,

Table 8.2 Items of the baseline survey

Areas	Information surveyed
Implementation of the WQM regulations	– Number of EISs (Environmental Impact Statement), POs (Permits to Operate), and certificates evaluated by each RO – Number of water bodies that have been classified – Status of pollution source control and surveillance for compliance – Status of self-monitoring report – Status of complaints and petitions from local residents on WQM issues – Guidelines and operational manuals other than the rules and regulations under the law
Organizational management	– Duties and responsibilities of WQM-related divisions in the ROs – Duties and responsibilities of staff in WQM-related divisions in the ROs – List of job requests from the EMB CO to the ROs; and the response of the ROs to the CO
Relationship with other organizations	– Activities in coordination with other agencies – PCO (pollution control officer); training plans and records
Physical resources	– Office and laboratory space – Office equipment (computers, printers, scanners) – Laboratory equipment – Maximum number of samples that can be analyzed – Actual number of samples analyzed in the last five years – List of analytical parameters that can be analyzed in-situ or in the WQM laboratory – Availability and status of field vehicles – Available maps
Human resources	– Number of staff – Staff qualifications – Reshuffling, status and the proportion of the staff turnover – Recruitment plan – Training and seminars (regular, ad hoc) – Status of staff training
Financial resources	– Annual budget allocation – Breakdown of the budget – Sources of earnings – Budgetary deficits/savings – Activities/items that require an additional budget

Continued

Table 8.2   Continued

Areas	Information surveyed
Information resources	– Number of stations and samplings per year in each water body – Numerical data regarding EIA-related activities in the last five years – Data management systems: Monitoring, processing, and storing data Dissemination of information to the public and/or related agencies Preparation of the capability report for the CO of the EMB

Source: Compiled from JICA and Philkoei International (2004).

the ROs lack experience in scientific field survey. Moreover, identification of the source of pollution is based only on the permit applications of companies.

Organizational management   The ineffective transformation of the EMB into a line bureau appears to be an issue. Although a DENR Administrative Order pertaining to the conversion of the EMB into a line bureau was issued in 2002, the staffing pattern and the structure of the bureau remained unchanged.The CO of the EMB tends to instruct the ROs and assign field operations to them without taking their capacity into consideration (Ohta and Kojima, 2003). As a result, ROs are unable to respond to the requests of the CO.

Human resources   The current manpower resources of the ROs indicate that the staff dedicated to WQM is only 9 per cent (76 staff members) of the PCD (Pollution Control Division) and RO personnel. Regions II, V and XIII have the lowest resources, with only two staff members assigned to do WQM work (JICA and EMB, 2005).

Financial resources   The main reason for selective enforcement of WQM tasks by the ROs is limited funding. With an average allocation of 2.5 per cent from the DENR budget over the last 11 years, it is not surprising that the ROs have been unable to implement fully the WQM mandates (JICA and EMB, 2005).

Physical resources (laboratories)   Except for ROs II and VIII, all ROs are capable of determining conventional water pollutants (BOD_5, DO, pH, TDS, and TSS). However, the RO laboratories lack the major equipment and accessories required to analyze organic pollutants and heavy metals. Moreover, most of the RO laboratories are incapable of conducting a bacteriological analysis of the water samples. In addition, on-site testing apparatus such as pH meters, conductivity meters, DO probes and turbidity meters is absent (JICA and EMB, 2005).

Information resources   All ROs can store and process WQ data but have adopted different practices in managing their databases. Most of them store their data as either printed copies or electronic files. These differences in database management practices, or the lack of a systematic WQ database at an RO, is mainly due to the lack of technical capacity and manpower, which in turn is caused by insufficient funds. Moreover, the ROs have repeatedly conveyed their difficulty in processing the huge amounts of WQ data accumulated over the years.

Required capacity

The following are the major elements of the CWA mandated to the EMB, and many of them, except water quality guidelines and effluent standards, are new mandates for the EMB (that is, mandates that the EMB has no experience in handling):

- Establishment of WQMAs and Water Quality Management Area Governing Boards (GBs)
- Designation of non-attainment areas and formulation and implementation of the necessary measures to improve water quality
- Coordination and provision of technical assistance to the appropriate agency for water supply and to sewerage facilities for the proper collection, treatment, and disposal of sewage
- Administration of the Water Quality Management Fund
- Categorization of the industry sector for establishing effluent standards specific for each industry
- Formulation and implementation of a waste water charge system in all water management areas
- Designing a discharge permit system applicable to all dischargers
- Allocation of effluent quotas by developing procedures and guidelines related to the water quality guidelines and pollution loads
- Preparation of the National Water Status Report, Integrated WQM Framework and a ten-year WQM Action Plan
- Preparing and publishing ground water vulnerability maps and classifying ground water resources
- Review and revision of water quality guidelines

- Review and revision of effluent standards
- Categorization of point and non-point sources of water pollution
- Classification and reclassification of water bodies
- Promoting and encouraging the private sector to use water quality management systems.

Table 8.3 indicates the distribution of the CWA mandates between the CO and ROs.

These CWA mandates make it necessary to enhance the capacity of the EMB. The important points of required capacity are as follows:

WQM policy and activities

To implement the CWA, first, the EMB needs to develop guidelines and manuals for the above-mentioned points. These guidelines and manuals need to be developed by the staff at the EMB in consultation with outside experts, instead of being outsourced. Second, the following existing processes need to be enhanced to meet the additional requirements of the CWA (JICA and Daruma Technologies, 2005).

Water quality monitoring includes not only sampling and analyzing water pollutants, but also examining other properties of the water body in order to establish the carrying capacity of the water body. Moreover, water quality analysis has to include other parameters that are identified for each industry sector.

The process of granting permits should finally evolve from a quota to a market-based instrument, rather than be a pure command and control tool, and the permits should specify the permissible quantity and quality acceptable from dischargers.

With regard to creating public awareness about the implementation of the CWA, information, education, and communication (IEC) requires situational analysis or audience profiling to be conducted before designing the strategies and activities. Appropriate monitoring and evaluation should be conducted during the entire process. IEC materials and strategies should tap into people's feelings, thoughts, motives, values and learning styles.

Establishing WQMAs and support institutions

The ROs lack trained personnel who can act as technical secretariats for the Area Governing Board. Moreover, the ROs have only a limited experience in mobilizing public participation for preparing WQMA action

Table 8.3 Assignment of the CWA mandates to the CO and ROs of the EMB

	Central office	Regional offices
Policy and planning	– Develop the integrated WQM framework – Develop the guidelines to designate WQMAs and non-attainment areas – Develop the guidelines to develop WQMA action plan – Develop the guidelines for operation of the Governing Board, Technical Secretariat, and multi-sectoral group – Develop the guidelines to issue the discharge permit and collect the waste water charge – Develop the programmes for regional offices to classify the water bodies and the ground water sources – Develop the programmatic EIA – Develop the guidelines to improve the Regional environmental laboratories – Develop categories of industry sectors – Develop water quality variance for geothermal and oil and gas exploration	– Develop the policy thrusts and priorities of the regional offices in terms of designation of WQMAs, non-attainment areas, WQMA action plans, classification of water bodies and groundwater sources, etc. – Develop the programmes for classification or reclassification of water bodies – Develop the programme for water quality monitoring in classified water bodies – Develop the programmes for plant inspection
Research and scientific analysis	– Develop the National WQ Status Report – Develop the National ground water Vulnerability Map – Review and revise the water quality guidelines – Review and revise the effluent standards – Develop the official testing procedures and the accreditation system of environmental laboratories	– Develop the Regional WQ Status Report – Develop the Regional ground water Vulnerability Map – Develop the programmes for the multi-sectoral group – Develop the regional WQ database for the purpose of effective regional water quality management – Classify water bodies and ground water sources

Continued

Table 8.3 Continued

	Central office	Regional offices
	– Categorize point and non-point sources – Classify the ground water sources – Develop procedures and guidelines to relate current water quality guidelines with total pollution loading so that effluent quotas could be allocated	– Undertake field research work necessary for the above activities
Enforcement		– Designate WQMAs and non-attainment areas – Develop and implement the WQMA action plan – Issue the discharge permits – Collect the waste water charge – Receive the self-monitoring reports and verify them – Implement the clean-up operation, when required – Undertake the plant inspection – Operate the regional environmental laboratory for the purpose of enforcement
Coordination	– Develop the National Sewerage and Septage Management Programme – Develop the guidelines for its domestic sewage collection, treatment and disposal – Develop appropriate incentives for business and industries – Strengthen the linkage mechanism with DepEd, PCG, DA, DPWH, DOH, DOST, CHED, DILG, and PIA	– Implement the WQMA action plan together with other organization concerned – Implement the programme of water quality surveillance and monitoring for the multi-sector group with other organizations concerned – Strengthen coordination with local government units
Fund management	– Manage the National WQM Fund	– Manage the Area WQM Fund

Source: JICA and EMB (2005).

plans and in establishing partnerships with industries and stakeholders for water quality monitoring and management.

Organizational management

To implement the WQM policy procedures in an RO, the CO needs to enhance its capacity to lead and support ROs; these include setting priorities in each region and providing instructions on procedures, database, reporting system and training programmes.

Human resources

The CWA mandates require a dedicated staff to be working under the WQMS; therefore, the organizational structure of an RO, especially the WQMS units, needs to be strengthened.

For example, with regard to water quality monitoring, the average time spent in monitoring water quality is about 18 hours. Thus, the time required to monitor 421 principal rivers, 71 lakes, and 22,540 kilometers of coastline in the country is 44,514 person-hours. If we assume that a person works 40-hours a week for 52 weeks in a year 21 staff members are needed merely to monitor these water bodies once a year. Further, assuming that these 21 members will be dedicated solely to water quality monitoring, the rest of the staff will have to inspect 362 firms a year, which underscores the dire need to hire additional staff dedicated to WQM (JICA and EMB, 2005).

Another human resource concern that has been overlooked, and which has hampered the existing WQM performance, is the development of RO personnel. Personnel development should complement the number of personnel needed in the implementation of the CWA. The specific tasks prescribed by the CWA require advanced training and education in WQM, in addition to knowledge acquired through experience.

Financial resources

The main reason for selective enforcement of WQM tasks by ROs is limited funding. The initial CWA appropriation of 100 million pesos was to be taken from government savings, if any, until such time that the activities could be funded through the General Appropriations Act. However, as experienced under the Clean Air Act, these funds were not guaranteed, despite being provided for by law. For enforcing the CWA, external sources will be required. The National Water Quality Management Fund (NWQMF) is designed to fund activities mandated by the CWA and it will be sourced from permit fees, fines and donations. However, the

232

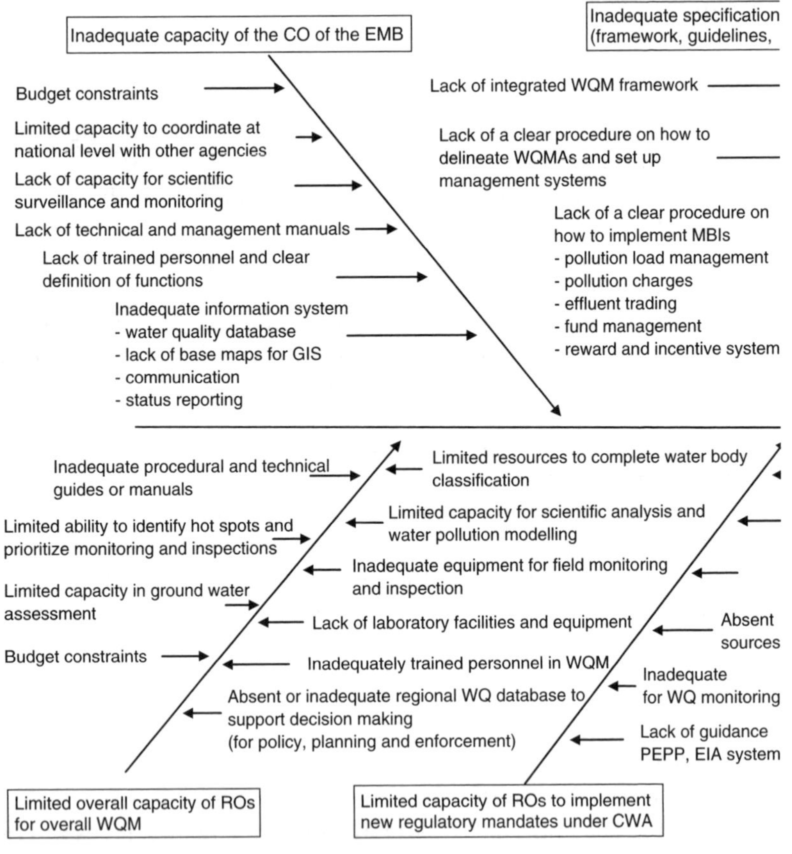

Figure 8.6 Capacity gaps in the CO and ROs

Source: Adapted from JICA and EMB (2005).

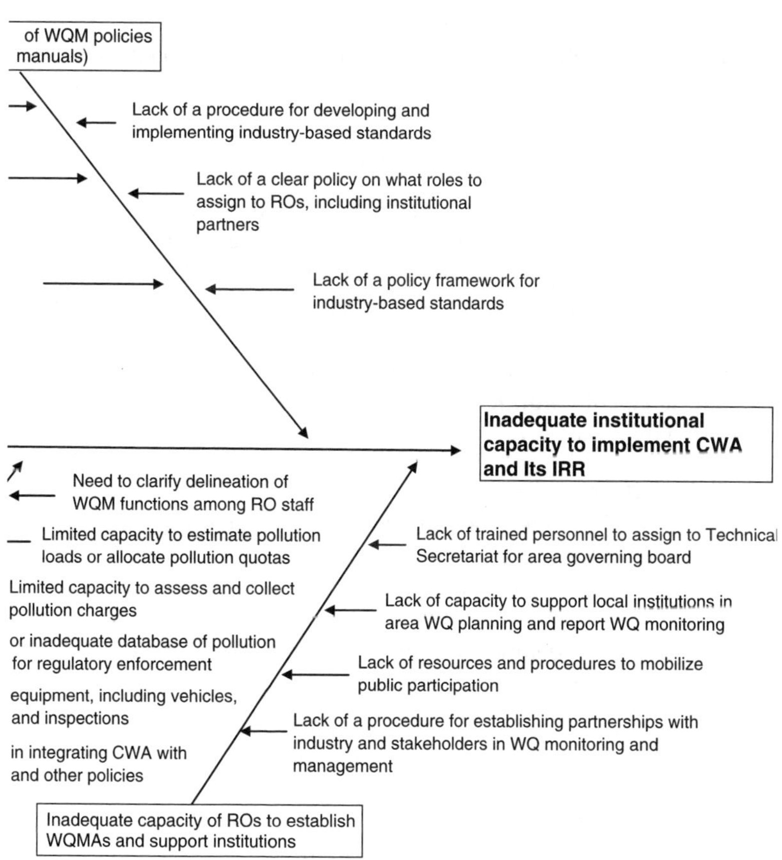

of WQM policies manuals)
Lack of a procedure for developing and implementing industry-based standards
Lack of a clear policy on what roles to assign to ROs, including institutional partners
Lack of a policy framework for industry-based standards
Inadequate institutional capacity to implement CWA and Its IRR
Need to clarify delineation of WQM functions among RO staff
Limited capacity to estimate pollution loads or allocate pollution quotas
Limited capacity to assess and collect pollution charges
or inadequate database of pollution for regulatory enforcement
equipment, including vehicles, and inspections
in integrating CWA with and other policies
Lack of trained personnel to assign to Technical Secretariat for area governing board
Lack of capacity to support local institutions in area WQ planning and report WQ monitoring
Lack of resources and procedures to mobilize public participation
Lack of a procedure for establishing partnerships with industry and stakeholders in WQ monitoring and management
Inadequate capacity of ROs to establish WQMAs and support institutions

NWQMF requires time to develop to levels that can sustain the CWA implementation.

Physical resources (laboratories)

A full inventory, along with inspection and verification of all equipment, is needed to identify the instruments that can still be used and those that need to be repaired or replaced. Through this exercise, the requirements for procurement and repair of equipment can be clarified.

Further, a new laboratory facility is required for some regions in order to accommodate equipments and to ensure secure and safe working conditions.

Information resources

Moreover, there is a need to develop a standardized water quality and pollution source database, which can be also used as a reporting system. The database can map pollution sources using GIS and the procedures for data input and utilization will be clarified, such as acquiring data from discharge permits.

These provide important baseline information for calculating the user's fee and, later, for confirming or validating the self-monitoring reports submitted by industries.

Capacity gap

The capacity gap is identified by comparing the current and required capacities. Since the focus is on the CWA requirement and organizational capacity, the major gap is described in the CO and ROs, as shown in Figure 8.6.

Taking into consideration the interrelation between the CO and RO, these gaps have been summarized into four major areas as follows:

- lack of an integrated WQM framework, including supporting procedures and guidelines for implementing such a framework, within the context and mandates of the CWA IRR
- inadequate capacity of the CO of the EMB to lead and support ROs in integrated WQM and CWA implementation
- inexperience and inadequate capacity of ROs in facilitating the establishment and operation of WQMAs, as well as their associated participatory systems and institutions
- overall lack of technical and management capacity among ROs in water quality management and, specifically, in implementing new regulatory mandates under the CWA IRR (that is, the discharge permit and waste water charge system).

Project design

Based on the gap shown (pp. 232–4), a project strategy was developed that specifically targeted the capacity of the EMB.

The goal set by the CWA can only be achieved over a long–term period and, during this term, there is an equally wide range of activities that need to be pursued. Although project assistance can possibly focus on only one or two selected activities that require support (for example, policy framework development and relevant guidelines), the impact of such an approach would not be significant unless there is equal support for the implementation of these policies and guidelines. On the other hand, if project assistance were to focus mainly on implementation activities (for example, regulatory enforcement activities in the regions), then the lack of a coordination framework, procedural guidelines and management tools would become serious obstacles (JICA and the EMB, 2005). Thus, the project broadly covers the requirements of the CWA and its goal reflects the overall required capacity.

Within the EMB itself, implementation of CWA mandates are divided between the CO and the ROs. Much of the support is needed in the regions that require CWA implementation activities, such as establishing area-based water quality management systems, regulatory enforcement and monitoring with the use of new instruments. In this case, apart from assisting the CO in preparing an integrated implementation framework and related procedures, the project will also strengthen the CO in becoming an effective supporter to ROs, where the policy and procedures will actually be applied. In addition, the project should focus on the role of the CO and ROs, and communication between the CO and ROs (JICA and EMB, 2005).

Project outputs are derived from the gap discussed (pp. 232–4) and the above strategy.

In addition, for selecting activities to be included in the project design, the following points were considered through a focus group discussion:

- Preparing procedures and conducting training for the implementation of the new tasks mandated by the CWA
- Providing a two-directional support, to the CO and ROs, and clarifying the role of the CO in leading and supporting the ROs
- Financial limitations
- Selecting three regions that will serve as test areas for applying the policy implementation procedures and management tools developed under the project.

Table 8.4 Project goal, output and activities

Goal, Output	Activity
Project Goal:	Development of national capacity to implement an integrated water quality management system within the context of the CWA
Output 1:	Integrated policy framework for WQM based on the CWA is established and supported by adequate procedural guidelines and training for EMB staff
	1.1 Set up multi-agency coordination system to formulate an integrated water quality management framework and implementation plan
	1.2 Prepare procedural guidelines for designating WQMAs (including the identification of non-attainment areas as defined under the CWA)
	1.3 Formulate a comprehensive policy on the use of market-based instruments for WQM, including procedural guidelines for implementation
	1.4 Prepare procedural guidelines for classifying inland and marine water bodies as well as ground water, including guidelines for conducting ground water vulnerability mapping
	1.5 Prepare procedural guidelines for facilitating WQMA action planning (by the Area Governing Board) and follow-on compliance planning local government units (LGUs)
	1.6 Prepare procedural guidelines, including system and procedures, for pollution load and charge computation in support of the discharge permit system
	1.7 Prepare procedural guidelines for managing the National WQM Fund
	1.8 Prepare procedural guidelines for the categorization of industries, including point and non-point sources of water pollution
	1.9 Develop an approach and prepare guidelines for establishing cooperation programmes with other agencies and civic groups in water quality monitoring
	1.10 Prepare guidelines and initiate coordination arrangements for allowing flexibility in enforcing discharge standards for specific types of industry sources

1.11 Prioritize pollution sources and prepare an operations manual on conducting compliance inspections for various types of polluting facilities
1.12 Review water quality guidelines to provide a basis for water reclassification and the revision of effluent standards
1.13 Design and implement a training programme for EMB CO and RO staff in all regions for each set of procedural guidelines, prepare training materials and conduct the training

Output 2: Capacity of EMB CO to lead and support the ROs is strengthened

2.1   Establish a coordination system with EMB ROs in implementing the guidelines developed under Output 1
2.2   Select or develop appropriate water quality modeling techniques, including calibration, testing and demonstration in selected regions
2.3   Design, develop, trial and implement a national information campaign for raising public awareness of water quality management issues
2.4   Design and develop a water quality and pollution source database management and reporting system for use by ROs, with capability for mapping pollution sources using GIS
2.5   Design and develop an Internet-based WQM information and communication system to link the EMB CO with the ROs
2.6   Integrate regional reports and publish the first National Status Report on water quality
2.7   Implement procedures for managing the National WQM Fund (based on procedural guidelines developed under Activity 1.7)
2.8   Procure sampling equipment for WQMS staff, and streamline operations of the EMB central lab as a reference laboratory and training centre for RO laboratory personnel
2.9   Design and implement a training programme for EMB CO staff on the use of the information and communication system developed, including fund management
2.10 Conduct activities to generate resources for non-pilot ROs; e.g., planning workshops with other donor agencies (e.g., World Bank, ADB)

Table 8.4 Continued

Goal, Output	Activity
Output 3:	Capability of EMB ROs to establish and support WQMAs and related institutions is strengthened in 3 pilot regions
3.1	Implement the guidelines for WQMA delineation
3.2	Set up the Governing Board and Technical Secretariat for the designated WQMAs
3.3	Facilitate the formulation of WQMA GB action plans and LGU compliance plans based on the guidelines developed under Activity 1.5
3.4	Assist WQMA GBs in establishing and managing the Area WQM Fund and the activities of multi-sectoral monitoring groups
3.5	Assist in establishing area-based cooperation arrangements in water quality monitoring based on the procedures developed under Activity 1.9
Output 4:	Overall capability of EMB ROs in WQM is strengthened in 3 pilot regions
4.1	Identify attainment and non-attainment areas based on the procedures developed under Activity 1.2
4.2	Classify or reclassify water bodies as needed based on the guidelines developed in Activities 1.4 and 1.12
4.3	Implement the discharge permit and waste water charge system based on procedures developed under Activity 1.6
4.4	Set up collection and accounting systems for permit fees and waste water charges
4.5	Conduct pollution source inventories and water quality field surveys
4.6	Apply the water quality model developed under Activity 2.2; for example, in allocating pollution quotas in non-attainment areas
4.7	Implement procedures (developed under Activities 1.8 and 1.11) for pollution source categorization, prioritization and compliance inspections
4.8	Manage the database of pollution sources and water quality data survey results, and link the regional database to the national database at the EMB CO
4.9	Procure equipment for sampling and analysis, and develop training materials to enhance capability of EMB regional laboratories; also assist ROs in initiating laboratory partnerships
4.10	Prepare and disseminate the first Regional Water Quality Status Reports
4.11	Design and implement a programme for RO staff in the non-pilot regions to visit and observe WQM procedures being implemented in the pilot regions

Source: JICA and EMB (2005).

The project design is presented in the form of goals, outputs and activities in Table 8.4. At the project implementation stage, regrouping of activities is planned for efficient implementation.

Further, with regard to the overall management level of the project described in Table 8.1, it can be said that the project is at a level between management of pollution and environmental management.

Issues ahead

This chapter describes the concept and methodology of capacity assessment and project design in an environmental management area and describes how project design can be developed through capacity assessment. Although this chapter focuses on capacity assessment and project design, the process and analysis of this stage are utilized in project implementation and evaluation. It is difficult to measure capacity using simple indicators since capacity measurement reflects many aspects. At the project design stage, it is important to set indicators or measurement methods that enable the monitoring of changes in capacity. There are, however, other issues that need to be considered.

In the case of a capacity development project in a broader environment/dimension, instead of a specific organization, the methodology employed to describe the capacity based on factors should be elaborated. This is because many factors influence capacity – institutional, political, financial, cultural and social – hence, the methodology used to identify the influence of these factors and their interrelation is important.

Further, a methodology for rapid capacity measurement in small projects is needed. Since a comprehensive view is required when assessing capacity, understanding the critical factors is important than simplifying them.

In addition, in order to elaborate the methodology, cases of capacity development projects need to be accumulated and their project designs need to be analyzed.

References

JICA (2004) *Capacity Development Handbook for JICA Staff: For Improving the Effectiveness and Sustainability of JICA's Assistance* (Tokyo: JICA).

JICA (2005a) *Approaches for Systematic Planning of Development Projects in Water Pollution Control* (in Japanese) (Tokyo: JICA).

JICA (2005b) *Supporting Capacity Development in Solid Waste Management in Developing Countries – Towards Improving the Solid Waste Management Capacity of an Entire Society* (Tokyo: JICA).

JICA and Daruma Technologies (2005) *Study for the Development of the Implementing Rules and Regulations of the Clean Water Act (Phase II)*, vols I and II (Manila: JICA).

JICA and EMB (2005) *Project Document: Capacity Development Project on Water Quality Management* (Manila: JICA).

JICA and Philkoei International (2004) *Baseline Survey on EMB Region Offices for JICA–EMB Capacity Development Project on Water Quality Management* (Manila: JICA).

Matsuoka Shunji (2006) 'Capacity Development and Social Capacity Assessment (SCA)', in *Social Capacity Development for Environmental Management and International Cooperation: 21st Century COE Program*, vol. 3, Hiroshima, Hiroshima University.

Ohta Masahiro and Kojima Hiroyuki (2003) 'Framework of Technical Cooperation Project on Water Quality Management' (draft) (in Japanese) (unpublished).

Sakiko Fukuda-Parr, Carlos Lopes and Khalid Malik (eds) (2002) *Capacity for Development: New Solutions to Old Problems* (New York: UNDP).

UNDP (1997) *General Guidelines for Capacity Assessment and Development* (UNDP) (http://mirror.undp.org/magnet/cdrb/GENGUID.htm)

World Bank (2003) *Philippines Environment Monitor 2003* (Manila: World Bank).

9
Social Capacity Development for the Protection of Endangered Species in Japan[1]

Takashi Matsumura

Introduction

The Okinawa Rail – *Gallirallus okinawae* – is an endangered bird species that is endemic to Yambaru, Okinawa Island, Japan (Photograph 9.1). In 1981, when this bird was reported as a newly discovered species, its total population was estimated to be approximately 2000. However, a series of surveys conducted thereafter indicated that both its habitat and population have declined. The most recent survey concluded that its population would soon be fewer than 1000. The Okinawa Rail is endemic to Yambaru; therefore, a decline in this species in Yambaru is equivalent to its complete extinction.

Ada is a beautiful hamlet with a population of approximately 250 people. It is located on the south-eastern side of Yambaru and along the southern border of the area inhabited by the Okinawa Rail. The Okinawa Rails play a part in the everyday life of the people of Yambaru. Children see these birds on their way to school, and the local people encounter them during their daily farming practices.

Over the past few years, the people of Ada have been working toward the preservation of and coexistence with the Okinawa Rail. Innovative activities have been undertaken that have achieved significant success. These activities include establishing a community rule for the sound raising of household cats and organizing joint mowing practices.

The purpose of this study is to introduce these activities and attempt to answer the following questions pertaining to the development process of these activities by applying the conceptual framework of the 'Social

Photograph 9.1   Okinawa Rail
Source: Michio Kinjo.

Capacity for Environmental Management' (SCEM) approach:

- How were the activities developed in Ada?
- Who are the key actors, and how were those actors formulated in the course of developing these activities?
- What were the roles of these individual actors, and how was the collaborative network among these actors constructed?

Based on a historical perspective, this study also attempts to identify the essential factors enabling the community of Ada to create and sustain its innovative activities for the conservation of the Okinawa Rail.

Challenges faced in sustainable development in Yambaru[2]

Yambaru is a mountainous stretch that occupies the northern region of Okinawa Main Island. It comprises three villages – Kunigami, Higashi and Ohgimi. Kunigami village is subdivided into 20 communities, one of which is Ada (Figure 9.1).

What are the challenges faced by these communities in sustainable development? The following sections briefly examine the present status of the three pillars of sustainable development: environmental, social and economic conditions in Yambaru.

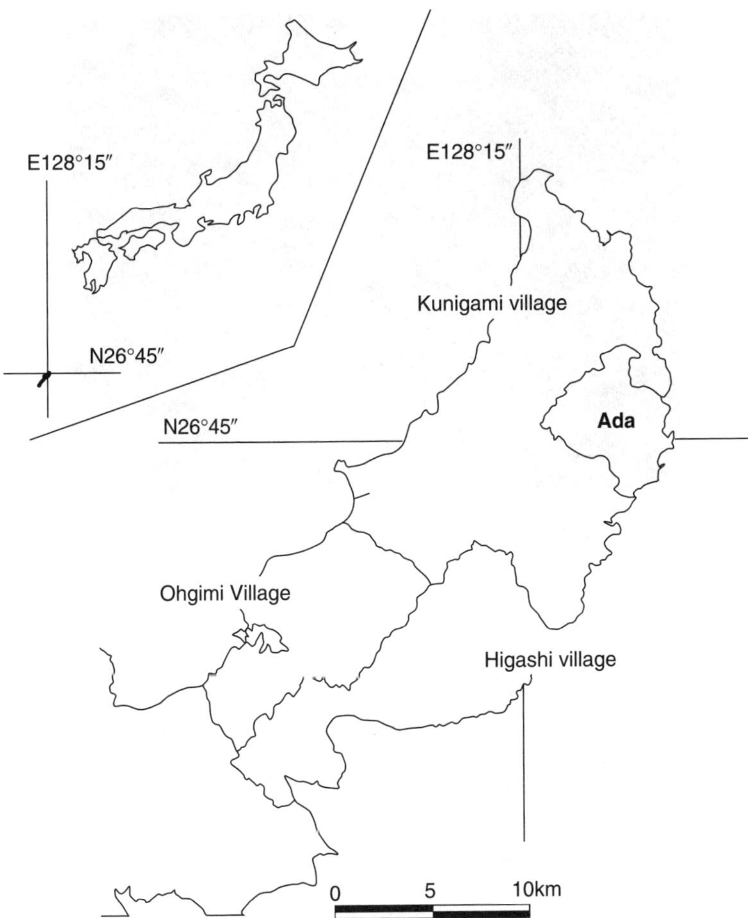

Figure 9.1   Yambaru area

Environmental conditions in Yambaru

Approximately 80 per cent of Yambaru is covered by subtropical broadleaf forests (Photograph 9.2). The communities of Yambaru, including Ada, are situated on the small plains along the streams that run in and between the forests.

The natural environment in Yambaru is characterized by its rich biodiversity. In addition to the Okinawa Rail, many rare bird species have

Photograph 9.2   Subtropical broadleaf forest in Yambaru
Source: Yasumasa Sawashi.

been reported to inhabit this area. Several rare mammals and insects have also been observed in this area (National Yambaru Wildlife Centre).

This forest and the inhabiting wildlife are sources of prosperity and blessing for the people of Yambaru. However, the natural environment does not necessarily benefit the local people all the time. Indeed, human lives have been jeopardized by Habu (*Trimeresurus flavoviridis*), a deadly poisonous snake. Its habitat is vast and spans the area, including the grasslands belonging to the communities. Although the number of Habu victims is decreasing every year, it poses a direct threat to human life and is the most serious concern of the local people.

There are no large-scale industrial activities taking place in this region; therefore, there is no hazardous industrial environmental pollution in this area. However, the implementation of various public works has led to the disposal of turbid water and sand; this is one of the most serious environmental problems faced by this region.

Solid waste is another problem caused by both local residents and visitors from outside the region. In fact, visitors from outside the region – such as tourists and anglers – are creating considerable problems. They often dump trash in the forest and along the seashore, which not only causes environmental contamination but also accounts for public hygiene problems.

An increase in the number of visitors has had another negative impact on the local environment. Some visitors abandon their pets, such as cats, in the forests. Such instances have been increasingly observed in Yambaru. This behaviour not only has an adverse effect on the daily life of the people, but also poses serious threats to the Okinawa Rail.

Socioeconomic conditions in Yambaru

The socioeconomic conditions in Yambaru can be summarized as depopulation, aging and a low income level. Over the past thirty years, there has been a steady decline in the population of Yambaru. The population of Kunigami village and Ada are presented in Table 9.1. It can be said that in the past thirty years, the population of Kunigami village and Ada declined by approximately 4 per cent and 23 per cent, respectively.

Table 9.1 Population changes in Kunigami and Ada

Year	Kunigami	Ada
1972	6,368	366
2000	5,991	247

Source: Ministry of the Environment (2005).

The tendency of the younger generation to migrate from the region invariably leads to an aging society. From among the entire population of 5900 villagers in the region, the proportion of villagers aged 65 and above is 26.8 per cent, as compared with the average figure of Okinawa prefecture being only 11.7 per cent. This indicates that the proportion of elderly population in Kunigami village is exceptionally high and increases with each passing year.

Although forest management has offered employment opportunities, the primary industry of Yambaru is agriculture. Further, the employment opportunities generated by forest management have reduced over the past few decades. The per capita income of the Yambaru region is presented in Table 9.2. Compared with the national average (approximately 2.91 million yen), the per capita income of Okinawa prefecture is low (accounting for 73 per cent of the per capita income at the national level). The figures for the three villages in the Yambaru region are even lower; they dropped to 58 per cent from 62 per cent of the national average.

Table 9.2   Per capita income

	National average	Okinawa	Kunigami	Ohgimi	Higashi
Per capita income (JPY 1,000)	2,910	2,125	1,805	1,673	1,802
Percentage	100	73.0	62.0	57.5	61.9

Source: Ministry of the Environment (2005).

Challenges to sustainable development in Yambaru

In view of the environmental and socioeconomic conditions in Yambaru, what are the challenges facing sustainable development?

First, it is of the utmost importance for the people of Yambaru to deal with the threats associated with Habu. Although the number of Habu victims has been decreasing, it is still evident that Habu pose a direct threat to human life.

Second, since the income level of Yambaru is lower than the average income level of the Okinawa prefecture, economic development seems to be a major cause of concern for the communities in this region with regard to realizing sustainable development. However, as far as Ada is concerned, this is not true.

People in Ada do not prioritize short-term economic development. In a recent interview survey conducted with the people of Ada, they expressed their deep concern for social conditions such as depopulation and aging. From their perspectives, depopulation and aging were more likely to lead the community to a loss of potential for future development (Daini Tokyo Bar Association, 2004). They were of the opinion that community development should be achieved by long-term efforts instead of short-term economic development, and that the younger generation and the existing rich natural resources provide the foundation for future development (Ikei and Shimabukuro, 2005; Matsumura, 2006).

Okinawa Rail's habitat

Habitat of the Okinawa Rail

The Okinawa Rail is an endangered bird species that is endemic to Yambaru, Okinawa Island, Japan. The first distribution survey on the Okinawa Rail was conducted in 1985, following which a series of other

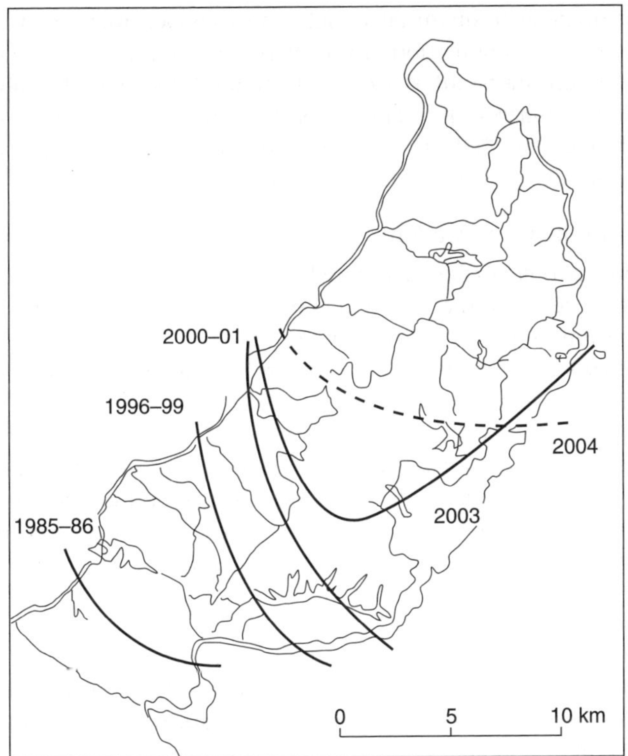

Figure 9.2   Changes in distribution areas of Okinawa Rail
Source: Ozaki (2005).

surveys were conducted. The results of these distribution surveys are summarized in Figure 9.2 (Ozaki, 2005).

Since 1985, there has been a steady decline in the habitat of these birds. Over the past 15 years, the southern border of the distribution area has shrunk by approximately 10 kilometres in the northern direction. Consequently, there was a 25 per cent reduction in its habitat area (Ozaki *et al.*, 2002; Ozaki, 2005). In 1985, the total population of the Okinawa Rail was estimated to be approximately 1,800 birds (Hanawa and Morishita, 1986). However, the most recent survey concluded that the total population of the Okinawa Rail would soon amount to fewer than 1,000 birds (Ministry of the Environment, 2005).

During this period, what were the conditions for the Okinawa Rail in Ada? Were there any changes in number with regard to its population?

There is no detailed quantitative data on this issue. However, it is quite unlikely that the population of the Okinawa Rail in Ada has decreased. In fact, according to interviews with the local residents, the birds are commonly observed on a daily basis. Moreover, since around 2000 or 2001, the Okinawa Rail has been observed more frequently in and around the community than ever before (Matsumura, 2006).

Threats to the Okinawa Rail

Generally, two factors are associated with the extinction of endangered wildlife – internal and external factors. The internal factors include the deterioration of genetic diversity within the concerned species and the size of its population. The external factors include the loss of habitat and an increased predation pressure.

Recent studies have revealed that external factors are identified as major threats to the Okinawa Rail; predation pressure from alien animals is a critical threat[3] as an external factor. Two of the alien species are identified as the major predators: the mongoose (*Herpestes javanicus*) and feral cats (*Felis catus*) (Ministry of the Environment, 2005; Ozaki, 2005).

Threats from the mongoose

In 1910, the mongoose was introduced as a predator of the Habu by the local people. Since their importation, the distributional area of the mongoose has been steadily expanding. Toward the end of the 1980s, their distributional area spread up to the southern border of Yambaru. Further, in the early 1990s, the mongoose spread towards the north. A recent survey concluded that the mongoose has gradually inhabited some of the forest trails within the Yambaru region. An increase in the habitat of the mongoose has led to an observed reduction in the habitat of the Okinawa Rail. It has also been observed that the Okinawa Rail disappeared from the areas inhabited by the mongoose. Therefore, it can be concluded that the mongoose is a threat to the Okinawa Rail (Ministry of the Environment, 2005).

Threats from feral cats

Another alien species that poses a threat to the Okinawa Rail is the feral cat. Feral cats are abandoned cats that used to be owned as pets. It has been observed that some of the visitors abandoned their domestic cats in and around the woodlands of Yambaru. These household cats rapidly multiplied in these natural surroundings. The local people of Yambaru own cats; the improper management of these cats also causes natural propagation. These feral cats are a grave threat to the Okinawa Rail.

At present, the increase in the number of mongooses cannot be directly attributed to human factors because their introduction was already banned. However, the problems associated with feral cats can still be closely attributed to human activities. Therefore, in addition to controlling the mongoose and feral cat population, the introduction of sound practices by cat owners can be regarded as an indispensable activity for the conservation of the Okinawa Rail.

Conservation of the Okinawa Rail in Ada[4]

In Yambaru, significant efforts have been made by various stakeholders to conserve the Okinawa Rail.[5] In Ada, the local people have been implementing a series of innovative activities at their own initiative.

The local children of Ada initiated activities to conserve the Okinawa Rail. These activities mobilized the local people and resulted in the establishment of a community rule to conserve the Okinawa Rail. This community rule was called the Regulation on Sound Raising of Cats. The regulation was enacted in May 2002 and was implemented with great success. However, this success created a new issue; namely, that the local people needed to develop a win–win solution for protection against the attacks by the Habu and conservation of the Okinawa Rail's habitat. In 2004, the people of Ada succeeded in developing a joint mowing plan as a win–win solution. These two activities are explained in the following section.

Children's activities and the Regulation on Sound Raising of Cats

Feral cats are one of the most critical threats to the Okinawa Rail. The primary source of these feral cats is their abandonment by visitors from outside the region. Moreover, domestic cats in the community that have not been soundly raised can be yet another source.

From the late 1990s until 2000, it was the children of the Yambaru region who initiated conservation activities. Various children's groups – such as the Children's Association in Ada – initiated activities to conserve the Okinawa Rail and to prevent the abandonment of feral cats. In order to conserve the Okinawa Rail, the children made posters and signboards and displayed them in the Yambaru area as an appeal against the abandonment of feral cats (Photograph 9.3).

These appeals by the local children changed the mindset of not only their families but also the other local people. These developments resulted in attempts to change their own activities and initiate the creation of a new community rule for the sound raising of pet cats.

Photograph 9.3   A signboard in Ada
Source: Yasumasa Sawashi.

A special committee was established under the Ada Community Council to review their daily activities and draft the rule. The initial challenge for the committee was to discover ways in which to sustain momentum and enforce innovation into the mainstream community agenda. To achieve this end, awareness raising activities – such as public symposiums – were organized by the committee. These activities appealed to the deeper consciousness of the people of Ada, to live in harmony with nature, and also provided them with the opportunity to redefine their everyday lives. Building on these activities, as well as based on the intensive discussions and consultations within the committee, the Regulation on Sound Raising of Cats was eventually enacted in 2002 in order to curtail the potential threats from feral cats to the Okinawa Rail.

This regulation does not include any penalties. In this sense, it should be regarded as a community rule. The rule comprises nine articles. The objective of the rule is explained in Article 1. It aims to ensure public health and conserve the natural environment of Ada by introducing necessary actions for the sound raising of household cats.

For this purpose, the authorities introduced a registration system by implanting microchips in the cats. This system enabled the community to easily identify their domestic cats. Under this rule, the local residents were requested to raise their household cats properly and refrain from abandoning them. In case this rule was violated, the head of the community would stipulate necessary recommendations to the residents concerned.

This rule was approved at one of the general sessions of the Residents' Assembly that was held in April 2002. This rule was put into effect in May of the same year. As a consequence of this rule, more rails were observed within the community. Frequent nesting and breeding of the rails were also reported in the grasslands in and around the community. Therefore, the new rule greatly contributed to the conservation of the Okinawa Rail.

Conflict of interests and a win–win solution

By controlling the abandonment of household cats, the community rule greatly contributed to the conservation of the Okinawa Rail. However, this success inadvertently paved way for another problem.

Ada abounds with grasslands. These grasslands are potential habitats for the Habu. Therefore, the local people used to mow these grasslands during early spring every year in order to minimize the risks from the Habu.

On the other hand, the Okinawa Rail inhabits these grasslands during their breeding season. Since the new community rule reduced the potential threats by feral cats, an increase in the number of Okinawa Rail and breeding activities were observed during the early spring season in 2003. During their mowing activities, the local people found numerous nests and Okinawa Rail fledglings more frequently than ever before. Therefore, the local people faced the problem of how to minimize the potential risks from the Habu, while simultaneously avoiding any disturbances to the breeding activities of the rails in the grasslands.

In order to resolve these two problems, in July 2003, the Ada Community decided to establish a new committee under the Council. The objective of this committee was to explore the possibilities of creating a win–win solution and reduce the risks from the Habu, while simultaneously conserving the rails. All possible solutions should entirely reflect the sentiments of the residents of the community. Moreover, they should be substantially formulated on a scientific basis–such as an evaluation of the potential risks from the Habu and an assessment of the possible impacts of mowing practices on the breeding activities of the rails. Therefore, in addition to the community councils, several external experts – supporting the councils – participated in the discussions.

In February 2004, the community councillors conducted an interview survey with all the residents of Ada in order to ascertain their sentiments. In parallel, they conducted a series of surveys on the habitat conditions of the Habu and the breeding activities of the Okinawa Rail. Extensive discussions and consultations were carried out at the committee in order to review traditional practices and formulate a new mowing plan.

A community-wide, joint working plan was formulated in order to equip the community to implement mowing practices in an effective and systematic manner. This plan identifies the location and areas of the sites that need to be mowed. This also includes the identification of the location of nests in order to protect the people from Habu attacks. Since joint mowing can disturb the nesting and breeding activities of the rails, a manual for the rescuing operation was prepared and distributed to the concerned people.

In March 2004, joint mowing practices were successfully implemented and completed with the participation of around fifty external volunteers. A follow-up survey revealed no significant changes in the nesting and breeding activities of the rails. In fact, after the implementation of joint mowing practices, a family of Okinawa Rails with fledglings was observed within the residential area of Ada (Ministry of the Environment, 2005). Moreover, there were no reports of Habu attacks (Matsumura, 2006).

Role of the individual social actors and their interrelationships

Innovation in Ada is a dynamic process and a chain of actions. What measures do the people of Ada undertake during the process of their innovative actions? The process of innovative actions comprises the following four stages: initiating specific actions, developing and implementing concrete actions, scaling-up the process by responding to changes and/or obstacles during the implementation of actions, and sustaining these innovative actions. As explained in the previous sections, the initial stage was led by the local children. While implementing the new community rule, the people of Ada faced a new problem and succeeded in formulating a win–win solution.

What conditions enabled the people of Ada to undertake innovative actions? What are the agents of change that influenced the community's transformation from one stage to another?

- Initial stage: From the late 1990s until 2000, the local children initiated actions to conserve the Okinawa Rail. What triggered such actions?

- Development and implementation stage: Between January and March 2002, the community rule was rapidly developed within a short period of time. How did this come about? Were there any specific factors accelerating this effort?
- Scaling-up stage: While implementing the regulation, the people of Ada encountered a conflict of interest in the form of the conservation of the Okinawa Rail versus protection from the Habu. How did they cope with these two conflicting requirements?

A framework of the historical review

In order to answer the above questions, the conceptual framework of the SCEM approach is applied. This approach assesses the capacity of a social system to manage environmental problems. According to the SCEM, a social system usually comprises three individual actors – the government, firms and citizens, and their interactions. Subsequently, a factor analysis is carried out to assess the capacity of the individual actors. The factors used in the analysis include those representing the institutional dimensions, human and other resources, and the knowledge base (Matsuoka, 2004; Matsuoka *et al.*, 2006). Since the SCEM approach was developed to assess the capacity of the social system primarily at the national level, it needed to be modified to assess the system at the community level, particularly in the case of Ada.

Primary social actors in Ada

The local people are the most important actors in the activities carried out in Ada. The logic behind this stipulation is that the local people themselves took the initiative in the form of a series of actions.

With regard to the government as an actor, among other organizations, the National Yambaru Wildlife Centre – a branch office of the national government – played a central role in the activities carried out in Ada.

In this case, who was the third actor? Under the SCEM approach, a firm, which is usually regarded as a primary actor, cannot be considered in the case of Ada. Instead, a group of experts comprising various professionals – such as veterinarians and scientists – assumed the role of the third primary actor. Under the SCEM approach, specialists or experts are regarded as an actor; however, they usually assume indirect roles. Therefore, they are assumed to be indirect contributors who provide information and/or technology to the primary actors such as the government, citizens and firms.

In the case of Ada, the experts, including veterinarians, not only accomplished their function of providing information and technology, but also played an even more important role. They were completely involved with the actual actions and played an indispensable role. In this respect, they can be considered as the third major actor. Contrarily, with regard to Ada, firms did not benefit directly from the conservation of the Okinawa Rail.

Factors influencing capacity

The local people, the governmental sector and the group of experts have been identified as the key social actors. Now, the question that arises is how their capacities can be assessed. What are the capacity factors in the case of Ada? In the following section, institutional, resource and knowledge dimensions are applied for a historical review of the situation. The factor analysis of the social capacities can be conducted quantitatively and/or qualitatively. However, it should be noted that the present analysis is restricted to a qualitative review.

Trigger for the children's actions

Generally, the most challenging task is to generate momentum to initiate innovative activities. In Ada, the local children assumed this responsibility. From the late 1990s until 2000, the local children generated the required momentum. However, the past few decades have not witnessed any remarkable and specific actions in Ada. What was the reason behind this? It is quite unlikely that the children had a first-hand experience in the decline in the rail population. Were any new children's groups formed? The answer to this question is 'No'. The Children's Association has an enduring history in Yambaru and no significant changes occurred in this regard. Were there any developments in other areas?

It can be indicated that there was a change in composition of the major actors. In fact, the National Yambaru Wildlife Centre was inaugurated and began operations in 1999. This centre was established by the Ministry of the Environment as the core centre for raising awareness, and the implementation of conservation and breeding projects. Further, comprehensive research and surveys for the conservation of the endangered species in Yambaru were conducted at this centre.

Immediately after its establishment, the centre, among other activities, began actively to collect information that was relevant to the conservation of the Okinawa Rail, and disseminated this information to the general public. The centre enthusiastically supplied information to visitors to the centre as well as the local residents of Yambaru. Further, by working in close collaboration with teachers, the centre organized a special programme to directly disseminate information to the children.

During the hands-on learning exercises of the programme, the children were deeply shocked to see the Okinawa Rail being killed in traffic accidents. After experiencing this, they realized the actual threats to the rails, recognized that such threats were closely associated with their daily lives, and realized the need to redefine their actions.

Important aspects of environmental education have been repeatedly emphasized, particularly hands-on learning. It can be concluded that the centre's efforts of information dissemination not only directly mobilized the local children to change their mindset, but also transformed these changes into actual action.

Factors accelerating rule making in Ada

In light of the action taken by the local children, the people of Ada began redefining their daily activities, and their efforts were accelerated in early 2002. Now, the question that should be asked is what were the factors accelerating this action?

First, there was a change in the composition of the social actors. In early 2002, the Association of Veterinarians to Protect the Okinawa Rail was established by the local veterinarians group. As a result, all the primary actors were formulated. During the formulation of the community rule, this newly established group of experts was totally involved as a practitioner. Based on its scientific knowledge, this group offered concrete options to the local people and implemented actions – such as infertility treatment for cats – in response to the requests of the local residents.

It was equally important that the National Yambaru Wildlife Centre acted as a catalyst while unifying the two actors – such as the expert group and the local people. In addition to its own responsibilities, the centre was actively involved in facilitating interactions among the social actors.

It should be noted that academic discoveries and media reports lent impetus to such movements. In 2001, during fieldwork carried out in Ada, an Okinawa Rail feather was found in the excrement of a feral cat. Despite the prevailing understanding about the negative effects of the feral cats on the Okinawa Rail, there was, until this discovery, no substantial scientific evidence supplementing it. This scientific discovery was sufficient evidence that the Okinawa Rail was actually threatened by the feral cats.

In December 2001, local newspapers carried numerous articles about this scientific discovery. These articles greatly accelerated discussions at the Committee of Ada. Further, the articles encouraged the local veterinarians to establish the Association of Veterinarians to Protect Okinawa Rail.

With regard to the factors of social capacity, the newly established committee considerably improved the institutional capacity of Ada. The committee offered fundamental opportunities for the direct exchange of views and ideas among the social actors. It also functioned as an effective tool in sustaining the momentum of innovation.

It should be noted here that Ada has a dynamic self-governing body, the Residents' Assembly, and all adults in the community have the right to attend the assembly meetings. The assembly is deeply rooted in the history of the community, and any regulations pertaining to local activities are subject to be discussed at this assembly. The Community Council, whose members are elected by the dwellers, is responsible for the management of day-to-day activities. Deeply rooted in history, the Residents' Assembly and the Community Council played the central role in community management.

A key to the success of the win–win solution

The regulation on the sound raising of cats was successfully enacted in May 2002. However, such a successful regulation resulted in a new conflict of interests: a trade-off between protection from the Habu and the conservation of the Okinawa Rail. However, by formulating the win–win solution, the people of Ada overcame this conflict.

How was this conflict resolved? In the course of formulating the regulation, three social actors and an effective collaborative network among them were formed. Was this enough to resolve the above conflict? Were there any changes in the composition of the social actors? If not, were there any changes in capacity factors?

First, the institutional capacity was changed by the establishment of the new committee on joint mowing. As was the case earlier, this special committee significantly contributed to the creation of the win–win solution.

While reviewing the capacity of individual actors, it was observed that there were changes in human resources as well; namely, new specialists such as reptile experts and ornithologists who joined the group of experts. These specialists had a considerable amount of experience and knowledge on the ecology of the Habu and the Okinawa Rail. These skills are essential to identify the potential risks from the Habu inhabiting the grasslands within the community. Further, the specialists were instrumental in identifying the distribution of the Okinawa Rail in Ada. With the participation of these experts, the committee was able to design a draft plan that was entirely grounded in science.

More importantly, there were significant changes in the knowledge base of the local people. In the course of discussions at the committee, action studies were introduced to identify the potential risks from the Habu and habitats of the Okinawa Rail within the community area. With the support from the newly joined experts, the local people themselves conducted these surveys. Utilizing these experiences, they accumulated their own knowledge instead of completely relying on the knowledge provided by the external supporting groups.

Essential elements enabling Ada to create and sustain its innovative activities

Although a large number of innovative initiatives are being implemented at the individual and project levels, it is difficult to encounter a genuinely creative and innovative community (Landry, 2006). It can be stated that Ada is an example of such an exceptionally successful community.

This leads to the following two questions. What are the fundamental factors that have enabled Ada to implement its innovative activities? Further, how can Ada sustain these activities? This section attempts to identify the essential factors that have enabled Ada to implement and sustain its innovative activities.

Building on the historical review discussed previously, the following five elements are identified as key factors responsible for fostering the innovative activities:

- Traditional self-governing bodies
- A long term and shared vision for community development
- Perfect harmony between the leadership and coordinating functions
- Empowerment of the community and the role of action studies
- Enhancing interactions and the role of the media as the fourth actor

Although these factors are distinct, they are closely interrelated. It should also be noted that all these factors hold equal importance.

Traditional self-governing bodies

Innovative activities tend to focus on the introduction of new and advanced technologies. Innovations by the community are likely to be more successful if the existing cultures, local knowledge and indigenous practices are respected and utilized. A recent study indicated that among other systems, the traditional governing system can be an effective tool for successful innovation at the community level (UNU, 2005).

Ada has self-governing systems such as the Residents' Assembly and the Community Council that are deeply rooted in the history of the place. As previously explained, such bodies have played a central role in the formulation and implementation of the new community rule. These organizations were also functioning as the ideal platform for the exchange of views and ideas for formulating the win–win solution and played a major role in the implementation of joint mowing practices.

It has been reiterated that ownership is essential to ensure the sustainability of activities. Without the ownership of a body concerned, activities can neither be effectively formulated nor successfully implemented. From the institutional point of view, maximum utilization of the traditional governing system is necessary for ensuring ownership.

Therefore, in Ada, the existence of such a self-governing system appears to be the key factor for the implementation of successful and sustainable activities, and ensuring ownership.

A long-term and shared vision for community development

Ada has envisioned a long-term vision for its sustainable development; this vision is shared among the people of the community. Such a clear vision plays a central role in pacifying the conflict of interests and generating a consensus among the local people. Further, it leads the people to implement concrete actions. For example, with regard to traditional mowing practices, there existed two contradictory opinions in Ada. Some of the residents were of the opinion that continuing with their traditional mowing practices would reduce the risks from the Habu; however, others believed that the traditional practices should be revisited. It was because of the long-term vision shared among the community that the people of Ada could manage to resolve such contradictory opinions and implement their innovative activities.

This vision appears to be indispensable even within self-governing bodies. Despite organizational changes, such as the alternation of leaders, a long-term vision enabled sustainable approaches. For instance, during the period between 2001 and 2006, the heads of the Ada Community Council were changed twice; owing to the expiration of his term, the head of the Community Council resigned. During the spring of 2002, a new head was elected at a time when the draft of the regulation was being finalized. As explained in the previous sections, these changes did not have any repercussions on the development and implementation of the activities within the community.

Therefore, it can be concluded that a long-term vision for future development is indispensable for creating a sustainable community. Also, it should be noted that such a vision must be shared with the people concerned.

Perfect harmony of leadership and coordinating functions

It is emphasized that leadership is extremely important for the implementation of innovative actions at community level. A recent study revealed that innovation is often initiated and implemented under the strong leadership of certain determined individuals (UNU, 2005; Yashiro, 2005).

The case of Ada is no exception. In fact, several outstanding leaders can be identified in Ada. With the inspiring ideas and excellent guidance provided by these leaders, the people of Ada were able successfully to design and implement their innovative activities. As an example, the local leaders assumed several critical roles, such as integrating the opinions of the local children into the mainstream community agenda, clarifying long-term visions and mobilizing the community people to participate in the discussions at the newly established committees under the Community Council.

The historical review of the case of Ada ascertains that coordinating functions are equally important for the development of innovative activities. The National Yambaru Wildlife Centre has been actively involved as a catalyst in order to integrate the activities of the local people of Ada and the group of experts. Through these efforts, the experts were able to have a clear understanding of what was expected from them. Accordingly, their scientific knowledge and skills were effectively integrated into the discussions held at the committee. Therefore, a perfect harmony of the leadership and coordinating functions – instead of the leaders alone – is important for the successful implementation of the activities at the community level.

Empowerment of the community and the role of action studies

New knowledge and skills are needed in order to initiate any innovative actions. These knowledge and skills are sometimes provided by the support of external professionals. However, this is not necessarily true in all cases.

Over the past few years, the conventional attitude of these external professionals – the 'we know best' attitude – has been questioned. As a result, there was a need for a change in their roles (Warburton and

Yoshimura, 2005). What a community requires are the knowledge and skills to solve the problems faced and not merely academic solutions. In this regard, it is necessary to apply the existing knowledge and skills to solving the actual problems. Further, a sound knowledge and deep understanding of these problems and the situations therein are indispensable for such an application. Therefore, the community should be empowered as the central actor. Hence, the community should accumulate its own experiences and knowledge instead of relying completely on external sources. In this respect, the action studies conducted in Ada in 2004 provided the local people with numerous opportunities.

As mentioned earlier, ownership is necessary for any environmental management activity. From the viewpoint of quality, the capability of the community itself should be developed. Without adequate knowledge and skills, the community cannot develop and implement any environmental management activities.

Encouraging interactions and the role of the media as the fourth actor

Since environmental management is a multidisciplinary issue, active interactions among the conservationists are indispensable for the smooth working of any environment management activity. Compartmentalization among the conservationists is very likely to occur during both the planning and implementation stages.

The case of Ada was no exception. Although several experts had frequently visited Ada, it is quite likely that they focused on their own professional interests. While the local veterinarians had been eager to do their bit to conserve the Okinawa Rail, they were unable to identify the aspirations of the local people.

This barrier was removed by the National Yambaru Wildlife Centre. Based on the historical review of the development process, it can be stated that there is another actor responsible for linking the individual social actors. This actor is the media. In the case of Ada, the local newspapers and TVs were actively reporting the situation in this region. These activities were quite helpful in terms of encouraging the local people and external experts to join the people's efforts in Ada.

As observed in the case of Ada, an environment management activity is a dynamic process. In such cases, it is important for the people from the media to provide the masses with new information. Further, since Ada is located in a geographically remote area, the role of media was particularly crucial for bridging the information gaps.

Conclusion

This chapter presents a case study on the development of the SCEM at community level. It introduces the people-led activities in Ada, Okinawa Island. These activities were aimed at conserving a globally endangered bird species – the Okinawa Rail.

By the application of the SCEM approach, the historical development of the creation of major actors and their effective interactions was reviewed. Further, the development of the capacities of individual actors was examined.

Building on the historical review of the development of activities in Ada, the following five elements were identified as the key factors responsible for fostering the initiatives of the local people: the existence of a traditional self-governing body; a long-term and shared vision for community development; perfect harmony of leadership and coordinating functions; empowerment of the community and the role of action studies; and encouraging interactions and the role of media. Although these factors are distinct, they are closely interrelated. Further, it should be noted that all these factors are of equal importance.

Environmental management at community level is a dynamic process. By the application of the SCEM approach, this chapter attempts to identify the agents of change through a historical review of the region under consideration. However, further analysis is needed in order to focus on the transition stages for a deeper understanding of the dynamism of the community.

Appendix

Interview survey

The interview survey was conducted by the author in order to obtain detailed information regarding the developmental process of the innovative activities in Ada. Further, the survey was designed to learn the opinions of the major actors on the agents of change and essential components. The interview was conducted during November 2005–July 2006.

In this survey, a total of 17 people were interviewed. These comprised five local residents, two local veterinarians, four governmental representatives, one employee of a private company and five experts from within and outside Ada.

Actions taken by the national and prefectural governments

The Okinawa Rail is registered as a Type 1A endangered bird species in the National Red Data Book. Further, it is registered as an endangered bird species by the Okinawa prefectural government.

In 2004, the National Protection and Breeding Plan for the Okinawa Rail was formulated by the national government. This plan calls for comprehensive action pertaining to conservation activities in and outside the habitat of the rail. These actions have been partially implemented.

Notes

1   The views expressed in this chapter are those of the author and not necessarily reflect the views of the United Nations University.
2   The sections on pp. 242–9 are prepared primarily on the basis of the report of the Ministry of the Environment (2005).
3   Road accidents are another threat to the Okinawa Rail. As far as the recorded deaths are concerned, road accident is the single largest cause (Kotaka and Sawashi, 2004).
4   The sections on pp. 249–61 are prepared on the basis of the unpublished interview survey conducted by the author (see p. 261).
5   The Okinawa Rail is registered as a Type 1A endangered bird species in the National Red Data Book. Significant efforts by the national and prefectural governments are summarized on p. 262.

References

Ada Community (2006) 'The Details of Activities in Ada-ku Ward', paper presented at the Population Viability Analysis Workshop on Okinawa Rail, January 13–14, Ada, Kunigami, Okinawa, Japan (in Japanese).

Daini Tokyo Bar Association (2004) 'Report of Interview Survey in Yambaru of Okinawa' (Tokyo: the Pollution Control and Environmental Conservation Committee of Daini Tokyo Bar Association) (in Japanese).

Hanawa, S. and E. Morishita (1986) 'Report of Rare Birds Survey in Japan', Tokyo, Environment Agency: 43–61 (in Japanese).

Ikei, T. and T. Shimabukuro (2005) 'Coexistence with Nature, and Conservation and Utilization of Local Resources', *Shimatati*, Public Association for Construction Services, Okinawa, Japan, 34: 16–18 (in Japanese).

Kotaka, N. and Y. Sawashi (2004) 'The Roadkill of the Okinawa Rail (*Gallirallus okinawae*),' *Journal of Yamashina Institute for Ornithology*, 35: 134–43 (in Japanese with English abstract).

Landry, C. (2006) 'Being Innovative, Against the Odds', in *Innovative Communities: People-centered Approach to Environmental Management in the Asia-Pacific Region*. J. Velasquez, M. Yashiro, S. Yoshimura, and I. Ono (eds), United Nations University: 46–79.

Matsumura, T. (2006) A personal communication with the staff of National Yambaru Wildlife Centre.

Matsuoka, S. (ed.) (2004) *International Development Studies* (Tokyo: Toyo-Keizai-Shinpo-sha) (in Japanese).

Matsuoka, S., K. Murakami, N. Aoyama, Y. Takahashi and K. Tanaka (2006) 'Capacity Development and Social Capacity Assessment', Social Capacity Development for Environmental Management and International Cooperation: 21 Century COE Programme 3, Hiroshima University: 1–24.

Ministry of the Environment (2005) 'Report of Model Project on Regional Development in Yambaru' (in Japanese).

Nagamine T. (2003) 'An Effort of Veterinarians for Conserving the Okinawa Rail', *Journal of the Japan Veterinary Medical Association*, 56–5 (in Japanese).

Nagamine, T. (2005) 'Constructing the Future for Okinawa Rail', *Shimatati*, Public Association for Construction Services, Okinawa, Japan, 34: 12–15 (in Japanese).

National Yambaru Wild Centre, http://www.sizenken.biodic.go.jp (in Japanese).

Ozaki, K. (2005) 'What is happening to the Okinawa Rail?', *Shimatati*, Public Association for Construction Services, Okinawa, Japan, 34: 6–8 (in Japanese).

Ozaki, K., T. Baba, S. Komeda, M. Kinjyo, Y. Toguchi and T. Harato (2002) 'The Declining Distribution of the Okinawa Rail (Gallirallus okinawae)', *Journal of Yamashina Institute for Ornithology*, 34: 136–44 (in Japanese with English abstract).

Sawashi, Y. (2005) 'Yambaru and Conservation of the Forests in Kunigami Village', *Iden*, 59–3: 84–90 (in Japanese).

United Nations University (2005) 'Research Report on Innovative Communities: People-centered Approaches to Environmental Management in the Asia-Pacific Region' (Tokyo: United Nations University Press).

Warburton, D. and S. Yoshimura (2006) 'Local to Global Connection: Community Involvement and Sustainability', in *Innovative Communities: People-centered Approach to Environmental Management in the Asia-Pacific Region*, J. Velasquez, M. Yashiro, S. Yoshimura and I. Ono (eds), United Nations University: 19–45.

Yashiro, M. (2006) 'Critical Elements for Achieving Community Innovations', in *Innovative Communities: People-centered Approach to Environmental Management in the Asia-Pacific Region*, J. Velasquez, M. Yashiro, S. Yoshimura and I. Ono (eds), United Nations University: 309–33.

10

Social Capacity Development for Environmental Compliance in Indonesia

Hoetomo

Indonesian environmental policies, programmes and problems

Water management policies and programme

Since water is one of the natural resources required for any activity, the government pays serious attention to it by issuing laws and regulations, and launching various programmes of water management. For example, the policy on the laws for and regulations on water can be seen in the promulgation of Act No. 7 of 2004 concerning Water Resources Management, and Government Regulation No. 82 of 2002 concerning Water Quality Management and Water Pollution Control (Ministry of Environment, 2004).

To promote environmental compliance, the government launched a programme named Program Peringkat Perusahaan better known as PROPER (a company performance rating programme) which later became a well-known programme on environmental management, especially in the environmental compliance of industrial sectors.

(a) Water management policies

The content of the above-mentioned law emphasizes the following issues:

Coordination and sharing of management authority.

The coordination and sharing of environmental management authority are implemented through the decentralization of water resources management, and the establishment of the Water Resources Council at every central, provincial and municipal level of governments. Besides, the law accommodates the public role and participation in water resources management.

Conservation, utilization and disaster control of water resources.

Water resources conservation entails the protection,preservation and maintenance of water resources, management of water quality, and control of water pollution. Disaster control of water resources is conducted by means of an early warning approach, while prevention and damage recovery is implemented through a comprehensive disaster information and mitigation system. In this context, the utilization of water resources is managed by means of regulations concerning the conservation and exploitation of water resources.

Privatization and social function of water resources.

In principle, the regulations stipulate that the government has full control on water management. Although the laws allow the private sectors to have large-scale involvement in the exploitation of water resources by means of licensing mechanisms, the social function of water resources and environmental preservation should be taken into account.

Control of water pollution.

Water pollution control is conducted based on Government Regulation No. 82 of 2002 concerning water quality management and water pollution control.

(b) Water management programme

PROPER is a programme that classifies plural companies on the basis of criteria that reflect the companies' compliance with environmental regulations. The programme is an instrument with which to monitor environmental compliance, and it simultaneously promotes transparency by involving the public (in the format of a council, the public consists of representatives from academia, NGOs and international organizations) in the process of defining the companies' performance rating. The companies are rated based on their compliance performance with regard to environmental regulations and the rating of every company is disclosed openly to the public. The public is expected to express appreciation to companies that comply with regulations and criticism to those that do not.

PROPER categorizes companies and business activities into five levels of compliance, represented by the following colours:

Gold: those surpassing the environmental requirements and achieving zero emission
Green: those surpassing the environmental requirements
Blue: those complying with the minimum requirements

Red: those not complying with the minimum requirements but showing a significant effort to achieve it

Black: those neither complying with the minimum requirements nor making any efforts to control pollution and environmental damages.

Water demand and water pollution

Population growth and speedy acceleration of development lead to an increase in the amount of water supply for households, farming and industries. The national demand for water in 2003 was as much as 112,275 million cubic meters and, in 2020, it is predicted to increase to 127,707 million cubic meters (Ministry of Settlement and Regional Infrastructure, 2003).

There are various factors that may result in the dwindling of the quantity and quality of water resources. Deforestation in the upper river basins, and changes in the function of catchment areas; conversion of function of forest areas into non-forest cultivation areas; the disappearance of lakes, swamps, and ponds; and long dry seasons contribute to the degradation of the quality and quantity of water resources. In addition, the industrial discharges of liquid waste into rivers cause water pollution. All these factors positively reduce the ability of land to absorb and retain rainwater, while pollution results in the deterioration of water quality.

A comprehensive monitoring of 72 great dams in various places during the dry season up to August 2003 showed that 23 dams can be categorized as dry, 38 dams are below the normal water level and only 11 dams can be considered as having the normal water level. According to research in 2003, the islands of Java, Bali and Nusa Tenggara experienced a water deficit of as much as 13.4 billion cubic meters.

In addition, industries and domestic activities also contribute to the degradation of the quality of water since they involve the discharge of waste into the river. For example, five rivers – the Cisadane, Ciliwung, Citarum, Cileungsi and Cimanuk in Java – have been heavily polluted. E. coli bacteria alone in the water of these five rivers is 2000 per millimeter. Consequently, the water of these five rivers cannot be utilized as a source of drinking water (Ministry of Environment, 2005).

Environmental dispute settlement

Significant growth and development in Indonesia have resulted in placing increasing demands on the delicate environment and in depleting the limited natural resources. Given these facts, problems related to the

management of forest and water resources have been particularly difficult to solve. Environmental cases are most likely to have the following
characteristics that lead to potential disputes:

- the cases conflict with development
- they involve powerless victims
- they require scientific evidence
- they relate to political issues.

Given the magnitude of these characteristics, an Alternative Dispute Resolution (ADR) mechanism involving public participation can play an
important role in resolving the dispute or conflict. Meaningful participation in public policy formulation could lessen the gap between those
who are affected by public decisions and those who are accountable for
formulating public policy (Thomas, 1989).

When considering the purpose of involving the stakeholders in the
decision-making process, it is important to determine the appropriate
public participation mechanism in a specific situation. The key objective
for public participation and conflict management processes is to assure
those who are affected by or subject to public policy decisions that they
have opportunities to:

- gain better information on the nature of the case
- Gain knowledge of potential consequences of public policy decisions
- Participate in the formulation of the decisions in order to avoid or at
  least minimize negative impacts and to maximize the benefits of the
  decision.

Environmental dispute settlement under the Environmental Management Act (EMA), Number 23 of 1997

The EMA provides two alternative mechanisms for the public to challenge environmental violations: in-court settlement (litigation process)
or out-of-court settlements via the ADR mechanism.

Environmental dispute settlement through an in-court settlement

In accordance with Article 34 of the EMA, for every activity that offends
the laws in the form of environmental pollution or damage, and causes
adverse effects on humans or the environment, the offenders are obliged
to pay compensation and/or carry out particular actions to restore the
environmental damage.

The elucidation of Article 34 stipulates that the above-mentioned article is a means of implementing the 'polluter pays' principle. In addition to the obligation to pay compensation, the judge may charge the offenders with an order to take a particular legal action; for instance, to install or repair a waste treatment facility or restore environmental functions.

Environmental dispute settlement through ADR (out-of-court settlement)

ADR is based on the voluntary choice of the parties in dispute. ADR is applied only to civil cases, and is never applied to criminal environmental cases. The purpose of ADR is for the parties in dispute to reach an agreement about the amount of compensation and/or to agree upon certain actions that need to be taken by the company responsible (ICEL, 1999a). Another important aspect of the agreement is the assurance expressed by the company that negative impacts on the environment will not reoccur or that a certain activity will not be repeated (Article 31).

Under Article 33 of the EMA, the government and/or community can establish an environmental dispute settlement agency that is free and impartial in nature. The settlement of environmental cases through an out-of-court process is carried out voluntarily by all the interested parties; namely, the affected parties that have experienced losses and those that have caused losses, government agencies and possibly the other parties that are concerned with the environmental issues.

If needed to facilitate the course of the out-of-court settlement, the interested parties might invite the involvement of a neutral third party jointly assigned by the concerned parties.

The third party could be one of the following:

- a mediator who serves to facilitate the negotiation with no authority to make decisions that affect the interested parties. This jointly appointed third party is assumed to:

  - possess sufficient and relevant skills to carry out the mediation process
  - have no type of relation with either of the parties in dispute
  - have no interest in the process of discussion or the outcome

- an arbitrator who is entitled to make fixed and binding decisions about all the parties in dispute.

Development of ADR in Indonesia

The concept of the ADR mechanism implemented in environmental disputes in Indonesia was introduced since the promulgation of the Principle of Environmental Management Act No. 4 of 1982. At that time, a number of individuals, organizations and government agencies in Indonesia had initiated the development of new collaborative systems to resolve a wide range of environmental disputes. However, until the late 1980s and early 1990s, the provision for public involvement and the implementation of ADR were still in the stage of concept development and had not formally begun (Takdir, 1996).

In the early 1990s, Professor Emil Salim, the then Minister of the Environment and Head of the Environmental Impact Management Agency (BAPEDAL) of the Republic of Indonesia, was committed to exploring ways in which ADR could be used to resolve specific contentious and difficult disputes. In 1990, a difficult dispute case, better known as the Tapak Case, took place in the city of Semarang, the province of central Java. The case pertained to the environmental pollution of a river brought about by ten companies located on the river bank; this adversely impacted the rice fields of the surrounding community. Nabiel Makarim, the then Deputy Minister for Pollution Control, Ministry of the Environment, was persuaded by the parties to conduct mediation as a means of resolving the contested issues. In this case, Lembaga Bantuan Hukum (an Indonesian legal aide) was appointed as the representative of the affected parties.

After a series of meetings (chaired by the mayor of the city of Semarang) conducted in the format of both face-to-face and private sessions, the parties reached an acceptable settlement. The accord contained the provision of financial contribution by the companies to the local communities, the development of an environmental protection programme, and pollution prevention initiatives. Irrespective of certain problems in implementing the pollution prevention and protection programme according to the agenda agreed upon, the use of the mediation mechanism was demonstrated as an effective mechanism for resolution of the dispute settlement among the parties concerned. It encouraged both the government agencies and environmental NGOs jointly to choose the type of mechanism to settle the dispute. Furthermore, the Tapak Case encouraged both the government and the NGO community to explore the system of mediation to resolve other environmental disputes and consider this mechanism as a successful model. However, they realized that this would require a significant level of awareness as well as support from government agencies, the business sector, NGOs and affected communities.

Examples of the analysis of water pollution cases using ADR

The tables below present the analysis of two environmental cases (the Tapak River case and the Siak River case) using ADR. The case of the Tapak River is presented in Table 10.1 and that of the Siak River, located in the district of Perawang, the province of Riau, is shown in Table 10.2.

General analysis of environmental cases using ADR

Considering the nature of the case, the process, the parties' involvement and the result, the following points emerge:

1 There was a lack of institutionalized and collaborative procedures available to address complex environmental disputes. So far, the procedures had been conducted only on an ad hoc basis;
2 Government agencies and informal leaders had played important roles in convening the process and encouraging the parties in dispute to use mediation mechanisms. In a number of cases, they successfully served as mediators. However, in other cases the government agencies were reluctant to serve as mediators, and were unable to monitor or enforce negotiated agreements or enforce pollution laws. They had limited capabilities or a lack of authority with which to protect the environment. These conditions limited affected their willingness to enforce compliance (Edward, 1985);
3 The NGOs were highly enthusiastic to implement collaborative procedures and, in most cases, preferred to implement them as judicial alternatives (Kimberlee, 1994);
4 The citizen groups were generally not well organized, were unable to create an impact and did not have the technical expertise to participate. These conditions prevented them from standing on an equal footing with the parties at the time of the mediation procedures. They need assistance in establishing an organization as well as in serving as legal advocates;
5 The companies, as well as local government agencies were likely to participate in mediation because of a desire to promote their public image, and to avoid adverse publicity, public unrest or court action (Christopher, 1986);

6 There were considerable differences between the involved parties with regard to the appropriate substantive scope of the mediation. One of the debatable issues in the negotiation process was whether the value of compensation should also be addressed or whether the discussions should be limited only to the stopping of pollution and mitigation of the effects of environmental degradation;

7 In a number of cases, there were difficulties in obtaining adequate, accurate and conclusive technical information useful for all the parties involved in the mediation process;

8 In several cases, party members were unclear regarding the appropriate role of the mediator. Opinions and actual performance on the issues ranged from an impartial process orientation to being a substantive advisor or evaluator;

9 Once the agreements were settled, significant difficulties were frequently encountered in implementing them due to a lack of clarity of the specific terms of agreement. There was also lack of implementation, monitoring and renegotiation procedures;

10 Despite some obstacles and limitations, a number of government agencies and NGOs were both enthusiastic and supportive in terms of applying the collaborative procedures.

Lessons learned and recommendations

Based on the above analysis, there are some lessons learned and recommendations that might be taken into consideration in the implementation of the ADR mechanism.

Lessons learned

Given the magnitude of the problem as well as the size of the country, the efforts to design, gain approval for and implement a nationwide environmental dispute resolution system for a country such as Indonesia are not so easy to implement. The country's diverse population and geography, as well as the range of serious environmental problems, have led to enormous challenges in implementing a viable system. These lessons have emerged from the observations of government officials, the key local Indonesian NGOs and academicians (Achmad *et al.*, 1998).

Consensus building on a national level, especially where multiple agencies and organizations with crosscutting mandates and responsibilities are involved, requires extensive time with regard to the development and implementation of a new system such as ADR (James and Wondolleck, 1990). However, once a consensus has been established, it

Table 10.1 Analysis of the Tapak River case

Location	Type of case	Stakeholders	Mediator	Settlement issues
Semarang, Central Java	The impact of river pollution on rice fields	1 Tugurejo community 2 Ten companies located near the Tapak River	Local government of Semarang City	1 Compensation 2 Environmental rehabilitation 3 Stopping of pollution

Result/agreement	Implementation	Opportunities	Obstacles
1 Financial contribution: industries preferred using the term 'contribution' to 'compensation' since the latter could reflect that the industries had been proved as the polluters 2 Establishment of an environmental management and pollution control programme	1 Lack of implementation of pollution control 2 Implementation of the environmental management programme did not meet the schedule	1 Demonstrated the role of BAPEDAL (local environmental management agency) as a new institution (at that time) 2 Obtained support from the central government 3 Demonstrated commitment as the key to settling the dispute	1 The role of the local government was too dominant 2 Lack of technical skill to design the agreement, with regard to the details of the pollution control programme 3 Difficulty in developing integrated waste management since the owner of the industrial estate was not involved

Table 10.2 Analysis of the Siak River case

Location	Type of case	Stakeholders	Mediator	Problems
Perawang, Riau	River pollution caused by a company named PT. Indah Kiat Pulp and Paper (PT. IKPP)	1 One hundread and fifty-seven households in Perawang 2 PT. IKPP	1 Deputy Minister for Pollution Control, BAPEDAL 2 Private sector consultant jointly appointed by BAPEDAL, the affected community and the company	The community experienced economic loss and suffered health problems due to water pollution

Result	Implementation	Opportunities	Obstacles	Notes
Development of the following programmes: 1 Environment management 2 Waste minimization 3 Community development	1 The community development programme did not run well because the budget was not sufficiently allocated and properly distributed 2 The communication forum between the community, company, and government was not realized as it should have been	1 Promoted the role of BAPEDAL as a facilitator in the mediation 2 Demonstrated the relationship between the government agency (BAPEDAL) and the NGOs 3 Pressure from the community and the NGOs was expressed through boycott and media press releases 4 Promoted the involvement of local NGOs 5 Gained support from the informal leader of the community in order to address the affected people	1 The NGOs and the affected community were unable to continue exerting constant public pressure 2 Difficulty in appointing an independent mediator 3 Not all the stakeholders were being involved (the local government was not involved)	The stakeholders reached the following agreements: 1 In principles, PT. IKPP agreed to control its pollution 2 The detailed agreement dealt only with a community development programme and did not include a pollution control programme

would affect new policies, structures, and procedures and make them much easier to implement as a high level of ownership and support will have been developed through the deliberative process.

Some lessons derived from the ADR mechanism are as follows:

1 The consideration of a consensus process to develop collaborative dispute resolution processes affects the enhancement of public participation. It is interesting to note that the development and greater use of collaborative procedures in Indonesia, at least at the local level, provides a genuine and, to a large extent, unique opening for broader democratic participation in crucial public issues. The actual cases that were mediated provided opportunities for government officials, representatives of the private sector, affected parties and the concerned NGOs to work with each other to develop mutually acceptable solutions to common problems. The successful use of these procedures has created a model to broaden democratic participation in the solution of public issues, and has provided procedural approaches that may be extended to address other demanding problems in the political arena;

2 Implementation of ADR in Indonesia should take into consideration traditions and practices based on local dispute resolutions. Indonesia has numerous cultural traditions in the consensus-building process. Indonesians in general, place a high value on building and preserving social harmony. The Indonesian decision-making tradition of *musyawarah* and the political philosophy of Pancasila (five principles) have supported the concept of building social consensus on important issues. Generally, people want to solve problems in ways that avoid social conflict (Suparto, 1999).

Recommendations

Considering the lessons learned, the following recommendations can be taken into account:

1 The need to clarify the differences between the role of convener, which can be assumed and played by the government, and the role of mediator, which must be played by a neutral and impartial service provider;

2 The effectiveness of ADR will depend upon the effectiveness of environmental law enforcement in Indonesia. Therefore, ADR for environmental issues must be developed simultaneously with strong, effective and consistent environmental law enforcement;

3 The establishment of public complaints management services along with the development of techniques and procedure in the environmental field would promote an effective ADR mechanism.

Steps to overcome shortcomings in environmental management

Considering that the success of environmental management relies on the capacity of social capital – namely, the government, the business community and citizens – the strategies entail the following:

1 the facilitation of law and regulation provisions, human resource development and institutional arrangement;
2 the establishment of a joint effort among the government institutions to launch regulations that are environmentally friendly. As an example, the Ministry of the Environment has a Memorandum of Understanding with Bank Indonesia (the Indonesian Central Bank) that, among other things, stipulates information exchange and the development of rules and regulations that are 'environmentally friendly'. As a result, one article of the Indonesian Central Bank Regulation stipulates that a debtor or recipient of bank credit has to manage the environment in its business activity. Poor environmental management could result in the increase of loan interest whereas good environmental management could result in the debtor enjoying a lower rate of interest;
3 the increasing of environmental awareness among the business community by persuading the latter to not only focus on economic profit, but also take into account the environment preservation in their activities. In this context, the government encourages the business community to participate in Corporate Social Responsibility (CSR) programs. As an incentive program, the Ministry of Environment in cooperation with Indonesia Accountant Association began conducting a CSR award program. The award is given to companies on the basis of their financial statements and efforts toward community development as stated in their annual report.

References

Achmad, S., T. Rahmadi and S. Migrantz (1998) *Environmental Mediation in Indonesia: An Experience* (Sangkar Madu, Jakarta: ICEL).

Christopher, W.M. (1986) *The Mediation Process, Practical Strategies for Resolving Conflict* (San Francisco: Jossey-Bass).

Edward, B. (1985) *Conflicts; A Better Way to Resolve Them* (Harmondsworth: Penguin Books).

ICEL (1999a) 'Alternative Dispute Resolution (ADR)', Working Paper for the Environmental Management Symposium, 10–13 April, Diponegoro University.

ICEL (1999b) 'Conflict Assessment Report; Case Study on PT Kayu Lapis Indonesia, Central Java and Mining Case on PT Newmont Minahasa' (Jakarta: Ministry of Environment).

James, C. and J.M. Wondolleck (1990) *Environmental Disputes* (Washington, DC: Island Press).

Kimberlee K. Kovach (1994) *Mediation Principles and Practice* (St Paul, Minesota: West Publishing Co.).

Ministry of Environment, Indonesia (2004) *State of the Environment in Indonesia 2004* (Jakarta: Ministry of Environment).

Ministry of Environment, Indonesia (2005) *State of the Environment in Indonesia 2005* (Jakarta: Ministry of Environment).

Ministry of Settlement and Regional Infrastructure (2003) 'Annual Report of Ministy of Settlement and Regional Infrastructure' (Jakarta: Ministry of Settlement and Regional Infrastructure).

Suparto, W. (1999) *Settlement of Environmental Disputes* (Surabaya: Airlangga University Press).

Takdir, R. (1996) 'Alternative Dispute Resolution (ADR) Mechanism', Law Enforcement Seminar on 1–4 January, Airlangga University, Surabaya.

Thomas, E.C. (1989) *Alternative Dispute Resolution; Melting the Lances and Dismounting the Steeds*, (Urbana and Chicago: University of Illinois Press).

Author Index

Subject Index